Balls Eight

History of the Boeing NB-52B Stratofortress Mothership

Brian Lockett

LockettBooks
www.Air-and-Space.com

Acknowledgements:

Boeing Archives:
Mike Lombardi
Air Force Flight Test Center History Office (AFFTC/HO):
Ray Puffer, Dr. James Young
Air Force Historic Research Agency:
Archie DiFante
NASA Dryden Flight Research Center:
Dill Hunley, Peter Merlin, Tony Landis, Dennis Ragsdale, Fred Johnsen, Bill Albrecht, Roy Bryant, Patti Peters
Bob Dorr, Greg Spahr, Terry Panopalis, Walt Boyne

Dedicated to my father,

Richard Lockett,

who inspired my interest in all things aeronautical.

About the author:

Brian Lockett has been a member of the American Aviation Historical Society since 1973 and earned his Private Pilot certificate as a teenager. He graduated from the University of California at Santa Barbara with a Bachelor of Arts degree in Geography. Brian worked for nearly a decade at the Santa Barbara County Air Pollution Control District. He is the creator of the Goleta Air & Space Museum web site (www.Air-and-Space.com). He is also the author of *Flying Aircraft Carriers of the USAF: Project FICON.*

.

Front cover: The NB-52B made its first public appearance at the May 17, 1960 Edwards Air Force Base open house with the X-15-1 mounted under its wing (Richard Lockett). NASA's Boeing NB-52B Stratofortress Mothership takes off on its last research flight with the third X-43A Hyper-X stack on November 16, 2004 (Brian Lockett).

Rear cover: *Composite of photographs of the NB-52B and the many research programs that it has supported over 45 years.*

LockettBooks
www.Air-and-Space.com

First Edition

ISBN 978-0-578-03000-5

Contents

Introduction

It has been asserted that the Boeing NB-52B Stratofortress, Air Force 52-0008 can lay claim to being the airplane that has seen and participated in more history than any other single airplane. For forty-five years, it has been a fixture at Edwards Air Force Base. The NB-52B and its sister ship, the NB-52A, 52-0003 were diverted to the mission of launching the North American X-15 hypersonic research airplane in 1958. The NB-52A was retired in October 1969, however the NB-52B continued to serve in the role of launch platform for a multitude of programs until its final mission on November 16, 2004.

The NB-52B was a flying launch pad, which is a highly complex task. It had to supply the rocket planes that it carried with the propellants, gases, and power normally associated with a launch pad at the Kennedy Space Center, all while flying nine miles above the ground. This ability made the airplane an indispensable asset for aeronautical research.

Its right wing was reinforced and equipped with a pylon to carry the heaviest wing-borne payloads ever dropped from an airplane. Some of the payloads that it dropped exceeded 50,000 pounds, nearly 1/5 of its own weight.

The X-15 rocket planes launched by this Stratofortress carried pilots to unprecedented speeds and altitudes. They were the first winged vehicles to reach speeds of Mach-4, Mach-5, and Mach-6. They were also the first winged vehicles to exceed 130,000 feet altitude, eventually reaching over 364,000 feet, high above nearly all the Earth's atmosphere. Five pilots qualified for astronaut's wings by exceeding an altitude of fifty miles in the X-15s. Tragically, one of those astronauts was killed on his qualifying flight when the third X-15 broke up as it re-entered the atmosphere.

The Space Shuttle program is indebted to the NB-52B in several ways. The modified Stratofortress launched a series of wingless lifting bodies that demonstrated the feasibility of the steep, gliding approach to landing used by the Space Shuttle. The NB-52B demonstrated that an airplane weighing a quarter of a million pounds could also perform such landings. The parachutes that lower the Shuttle's solid fuel boosters into the ocean were tested in drops from the pylon on its wing. The drag chute that slows the Space Shuttle after landing was tested on this airplane.

Despite repeated protestations that maintaining the antique bomber was becoming progressively more difficult and expensive, new programs kept emerging that required the unique capabilities of the NB-52B. As a result, it continued to haul unique devices aloft. It launched a variety of scale remotely piloted research vehicles and drones in the 1970s and 1980s. These vehicles included the 3/8-scale F-15 Remotely Piloted Research Vehicles (RPRV), a Ryan Firebee II drone, Ryan Firebee based Drones for Aeroelastic Structures Testing (DAST), and the Highly Maneuverable Aircraft Technology (HiMAT) RPRVs.

It was the last B-52 with a bomb bay configured to carry the MAU-12 bomb rack originally used to drop large thermonuclear warheads. That capability was utilized to drop a heavy device to evaluate improved parachutes for the crew escape module of the General Dynamics F-111 in a program that spanned thirteen years.

The NB-52B itself was used to conduct research. It helped researchers visualize wake turbulence from large aircraft by flying past a tower equipped with smoke generators. It served as an air-to-air gunnery target. For some time in the mid-1990s, the NB-52B flew as a ten-engine jet as it tested pollution reducing fuel additives with a pair of jet engines mounted under its bomb bay.

The NB-52B was the first airplane to launch a satellite into orbit. Orbital Sciences designed the Pegasus booster to be launched from its wing. The weight, dimensions, and payload capacity of the Pegasus were largely determined by the requirement that the rocket fit on the launch pylon. Following launch from the NB-52B, the Pegasus booster accelerated its payload to 18,000 miles per hour, more than three times as fast as the fastest X-15 flight.

The largest parafoil in history was dropped from the NB-52B during the X-38 program. The parafoil had an area of 7,500 square feet, greater than 1-1/2 times the area of the wing on a Boeing 747. It would have been used on the Crew Return Vehicle that was intended to provide a means of escape from the International Space Station.

The career of the NB-52B culminated in the launch of the fastest air-breathing jet engine in history. On its final mission it launched the third X-43A Hyper-X supersonic combustion ramjet on a modified Pegasus booster that propelled it to a speed of Mach-9.6 at an altitude of 110,000 feet. The successful operation of the supersonic combustion ramjet presages future aerospace vehicles that can cruise at over 7,000 miles per hour.

Due to the critical need for the unique capabilities of the NB-52B, it soldiered on far longer than its contemporaries. All other B-52Bs were retired in the 1960s and only the B-52H model remains in service with the Air Force. At the time of its retirement, it was the oldest flying B-52 by nearly ten years. Many of its parts and systems are completely different than those on the B-52H model still in service. It was powered by different jet engines that required an obsolete kind of jet fuel that was produced in special batches specifically for its use. It was kept flying by the adaptation of parts from nearly all the other models of B-52. The massive level of effort expended to keep it flying is a testament to its extraordinary capabilities.

The unique contributions of the NB-52B to aviation history were celebrated when it was honored with a retirement ceremony at the NASA Dryden Flight Research Center at Edwards Air Force Base on December 17, 2004. It will be preserved for posterity, on public display at Edwards Air Force Base.

The Origin of Air Launching

When the United States developed the first rocket-propelled,

Boeing B-29 Superfortress, 45-21800 was modified to carry the Bell X-1 rocket planes. (AFFTC/HO)

transonic research aircraft after World War II, it was believed that they would be operated in a conventional ground take-off mode. Both the Air Force Bell X-1 and the Navy Douglas D-558-II Skyrocket were initially intended to take off from the ground on their own power.

Reaction Motors produced the rocket engines used in the X-1 and Skyrocket. The X-1 used the XLR-11 and the Skyrocket used the nearly identical XLR-8. Each rocket was composed of four combustion chambers that could be individually toggled on and off. Each combustion chamber produced about 1,500 pounds of thrust, with a combined total thrust of about 6,000 pounds. The fuel used by these rocket engines was ethyl alcohol mixed with water.

The Bell X-1 was designed to use a turbopump to force the propellants into the rocket combustion chambers of the Reaction Motors XLR-11. The development of the turbopump required much more time than had originally been anticipated. In order to begin the rocket plane's flight program before the turbopump was ready, it was decided to install a high-pressure fuel tank system. While the design of the original propellant tanks closely matched the contours of the exterior of the X-1, the high-pressure tanks were spherical, with a greatly reduced volume. The burn time of the rocket engine was cut almost in half. The X-1 would not be able to take-off, climb to altitude, and reach supersonic velocity with the available fuel load. The performance of the resulting airplane would have been inadequate for the purpose of transonic testing.[1]

As a result of the delays in the development of the turbopump, it was necessary to make up the lost performance by carrying the X-1 to altitude under a large four-engine airplane. Consideration was given to using a Douglas C-54 Skymaster, but the end of World War II made a huge supply of Boeing B-29 Superfortresses available for the job. A B-29B, carrying Army Air Corps serial 45-21800 was diverted to Bell Aircraft in Niagara, New York to be modified for the purpose of launching the X-1.[2]

The use of a mother plane reduced the danger of getting the rocket planes off the ground. It also made glide tests of the flying characteristics of the airplane possible without requiring the rocket engine to be tested at the same time. On the other hand, it required the test pilot to learn how to fly and land the X-1 in just a few minutes as the it glided to Earth on his first flight.

The first air launch of a U. S. rocket research airplane took place at Pinecastle, Florida on January 25, 1946. B-29B, 45-21800 dropped the first X-1, 46-062 with Jack Woolams at the controls. Ten glide flights were made at Pinecastle. The last one was flown on March 6, 1946.[3]

The second X-1 carried serial 46-063. Chalmers "Slick" Goodlin flew it for the first time at Muroc Army Air Corps Base on October 11, 1946. He was launched on four glide flights of the second X-1 before making the first air-launched, rocket powered, transonic research flight on December 9, 1946. A total of eighteen contractor demonstration flights were made with the second X-1 at Muroc, the last being flown on May 29, 1947.[4] It was then turned over to the National Advisory Committee for Aeronautics (NACA).

Powered flights of the first X-1, 46-062 began on April 11, 1947. Goodlin flew a series of nine contractor demonstration flights between April 10 and June 5, 1947. 46-062 was then handed over to the Army Air Corps for envelope expansion flights. Captain Chuck Yeager achieved supersonic flight in the X-1 on October 14.[5]

The Douglas Skyrockets performed dozens of take-offs from Rogers Dry Lake. The first Skyrocket, carrying Bureau of Aeronautics number (BuNo) 37973 was powered only by a 3,000-

Boeing B-50A Superfortress, 46-006 was also modified to carry the Bell X-1 rocket planes. The X-1D exploded in the bomb bay of the Superfortress shortly after this picture was taken on August 22, 1951. (AFFTC/HO)

pound thrust Westinghouse J34 jet engine. Douglas pilot John F. Martin flew it for the first time on February 4, 1948. The second Skyrocket, BuNo 37974 made its first flight with NACA pilot Robert A. Champine at the controls on May 24, 1949.

The Skyrocket's take-off procedure involved running up the jet engine and releasing the brakes. The Skyrocket began its ground roll under jet power alone. As speed built up, the pilot ignited four solid fuel JATO units to kick the Skyrocket into the air. The JATO units were jettisoned after take-off.

The third Skyrocket, BuNo 37975 was the first to be equipped with the 6,000 pound thrust Reaction Motors XLR-8 rocket engine installed in addition to the J34 jet engine. On February 25, 1949, Douglas pilot William "Bill" Bridgeman made the first flight of the rocket-powered Skyrocket. It took off from the ground like its sister ships.

Eventually Boeing B-50A Superfortress, 46-006, equipped with four R-4360, 28-cylinder radial engines, was acquired to provide a second and more powerful launch plane for the X-1.

The first X-1 was retired after its last flight on May 12, 1950. The second X-1 made its last flight on October 23, 1951.

Using the operations of the Bell X-1 as their example, Douglas engineers concluded that air launching the Skyrocket would increase its top speed from Mach-1.08 to about Mach-1.6. The ability to operate in this speed range would allow the Skyrockets to generate swept wing performance data that complemented the data generated by the straight-winged X-1. In recognition of the apparent performance and safety benefits of air-launching, Hugh Dryden drafted a letter to Douglas proposing the modification

Boeing B-50A Superfortress, 46-011 was modified to carry the Bell X-2 rocket planes. The Superfortress was so severely damaged when X-2, 46-675 exploded in its bomb bay over Lake Erie on May 12, 1953 that it never flew again. (Bell Aircraft Corp.)

of the Skyrockets to air launch configuration and the acquisition of a P2B-1S (the Navy variant of the B-29) to act as the launch airplane.[6]

Captive carry flights of the third Skyrocket under P2B-1S, BuNo 84029 were begun in September 1950. The first air launch of the Skyrocket occurred on September 8. George Jansen piloted the P2B-1S. Bill Bridgeman was launched in the third Skyrocket over Rogers Dry Lake. The first few flights were made on jet power alone. Bridgeman flew the first air launched, rocket-propelled flight of the Skyrocket on November 17, 1950.[7]

The second Skyrocket, BuNo 37974 had its Westinghouse J34 jet engine removed and additional rocket propellant tanks installed when it was converted for air launching. It was first air launched on January 26, 1951. Bill Bridgeman made seven flights in the D-558-II #2 for Douglas. The Douglas demonstration program of the all-rocket Skyrocket ended on August 15, 1951. It was then turned over to NACA.

The first turbopump equipped X-1s finally arrived at Edwards Air Force Base in 1951. The third X-1, 46-064 had a similar configuration to the first two X-1s but much greater propellant tank volume. The X-1D, 48-1386 was a new design with a stretched fuselage and a more conventional cockpit canopy.

B-50A, 46-006 escaped destruction when the Bell X-1D exploded in its bomb bay on a propellant jettison test flight on August 22, 1951. It was not so fortunate when the X-1 #3 exploded in its bomb bay on the ground as Joe Cannon attempted to jettison its propellants after a failed propellant jettison test flight on November 9, 1951. The X-1 program did not resume flight operations until the delivery of the X-1A, 48-1384 in early 1953. B-29B, 45-21800 performed all subsequent X-1 launches.

Because the X-1 and the Skyrocket had been designed for take-off from the ground, they were equipped with conventional, retractable, wheeled landing gear. By the time the X-2 was conceived, the air launching of rocket planes was accepted as the standard method of operation. The X-2 was designed with a skid landing gear, which was unsuitable for take-off from the ground. Boeing B-50A Superfortress, 46-011A was acquired for the task of launching the X-2. The Superfortress was re-designated EB-50A following its modification as a launch plane.[8]

The first X-2 glide flight was flown at Edwards Air Force Base on June 27, 1952. Skip Ziegler was launched from EB-50A, 46-011A in the first X-2 that Bell delivered, 46-675 over Rogers Dry Lake.[9] X-2, 46-675 made three glide flights at Edwards Air Force Base and was then returned to Bell Aircraft in New York State for installation of its 15,000-pound thrust Reaction Motors XLR-25 rocket engine.

Bell made the first contractor demonstration flight of the X-1A, 48-1384 on February 14, 1953. Bell made six demonstration flights with the X-1A before turning it over to the Air Force in the spring of 1953.

X-2, 46-675 exploded in the bomb bay of EB-50A, 46-011A during a propellant jettison test over Lake Ontario on May 12, 1953. Skip Zeigler and Frank Wolko, a crewman aboard the Superfortress were killed in the explosion. The EB-50A landed safely, but it was damaged beyond repair and never flew again. A B-50D Superfortress, 48-096 was modified to launch the remaining X-2, 46-674.

Scott Crossfield piloted the all-rocket D-558-II #2 on the fastest flight of the Skyrocket program on November 20, 1953. It was the first flight to exceed a speed of Mach-2.

The Air Force made its first flight of the X-1A on November 21, 1953. Chuck Yeager piloted the X-1A to a maximum speed of

Boeing B-50D Superfortress, 48-096 replaced B-50A, 46-011 as the mothership for the Bell X-2 rocket planes. (Air Force Flight Test Center History Office)

Mach-2.44 on December 12, 1953. Major Arthur Murray piloted the X-1A to a maximum altitude of 90,440 feet on August 26, 1954. The Air Force made a total of eighteen X-1A flights before turning it over to NACA.

Bell delivered X-2, 46-674 to Edwards Air Force Base in 1954. Major Frank K. Everest piloted its first glide flight on August 5, 1954. Over the next two years, EB-50D, 48-096 launched it on four glide flights and thirteen powered flights.

Bell delivered the X-1B, 48-1835 to Edward Air Force Base in the summer of 1954. The Air Force had just concluded its X-1A research flight program and turned it over to NACA. B-29B, 45-21800 launched the X-1B on its first glide flight on September 24, 1954. The Air Force made two glide flights and ten rocket-propelled flights with the X-1B. Frank Everest reached a maximum speed of Mach-2.3 during the last Air Force X-1B flight on December 2, 1954. The X-1B was then turned over to NACA, which did not fly it until August 14, 1956.

NACA made one flight with the X-1A on July 20, 1955, but it exploded in the bomb bay of B-29B, 45-21800 before launch on its second NACA flight on August 8, 1955. The X-1A was jettisoned over the bombing range east of the base.

The second X-1, 46-673 was rebuilt with a turbopump, larger propellant tanks, and a more conventional cockpit canopy. In its new configuration it was designated X-1E. The X-1E was returned to Edwards Air Force Base in late 1955. B-29B, 45-21800 carried it on a captive carry flight on December 3, 1955. Joe Walker piloted its first glide flight on December 12 and its first rocket-powered flight on December 15. The X-1E made eight flights before Neil Armstrong flew the first NACA X-1B flight on August 14, 1956.

The third Douglas Skyrocket, BuNo 37975 made its last flight on August 28, 1956. It had flown twenty-one flights during the Douglas contractor demonstration program and sixty-six flights for NACA. NACA had directed Douglas to modify the first Skyrocket, BuNo 37973 to all-rocket, air launched configuration, but made only one flight with it on September 17, 1956 before retiring it.

X-2, 46-674 reached its maximum altitude of 126,200 feet with Captain Iven C. Kinchloe at the controls on its twelfth flight on September 7, 1956. Mel Apt piloted X-2, 46-674 to a speed of Mach-3.2 on its thirteenth powered flight on September 27, 1956. After engine burn out, it went out of control and Apt was killed when the cockpit escape module impacted the ground.

Jack McKay made the last Skyrocket flight in the all-rocket D-558-II #2, BuNo 37974 on December 20, 1956. NACA had flown it seventy-five times.

Rocket powered test flights continued with the Bell X-1B and X-1E. The X-1B made seventeen flights for NACA. Neil Armstrong piloted the last X-1B flight on January 23, 1958.

NACA had been incorporated into the National Aeronautics and Space Administration (NASA) by the time that John McKay piloted the last flight of the X-1E on November 6, 1958. It was the same airplane that had made the first X-1 glide flight at Muroc Army Air Field on October 11, 1946, and it was launched by the same B-29B. It had made twenty-six flights in its new configuration.

Sadly, the surviving Superfortress launch planes were retired to the boneyards. B-29B, 45-21800 and EB-50D, 48-096 were flown to the Military Aircraft Storage and Disposition Center at Davis-Monthan Air Force Base, Arizona and eventually scrapped. P2B-1S, BuNo 84029 was flown to the Navy storage facility at Litchfield Park, Arizona on August 15, 1959. It was restored to airworthy condition in 1969 and flown for a number of years. It is now awaiting restoration in the collection of the Weeks Air Museum in Polk City, Florida.

The Superfortresses that air launched the rocket-propelled Bell X-1, Douglas D-558-II Skyrocket, and Bell X-2 amply demonstrated the performance and safety benefits of the method during twelve years of supersonic research in the skies over the Mojave Desert. The first airplanes to fly Mach-1, Mach-2, and Mach-3 were launched by Superfortresses. As a result, the North American X-15 was designed from the beginning to be launched in flight.

Toward Mach-6

In November 1950, Hubert Drake and Bob Carmen, NACA's long range "dreamers" developed one of the first ideas for a vehicle capable of demonstrating flight at Mach-6. They envisioned launching a modification of the Bell X-2 from a mother plane equipped with five Viking rocket engines. The mother plane would have launched the research plane at 70,000 feet altitude and a speed of Mach-3. The research plane would then climb to Mach-6 at about 250,000 feet altitude, performance much like that eventually envisioned for the X-15. Walt Williams, the head of the NACA Muroc Flight Test Unit, determined that the report they submitted was "premature" and it was pigeonholed. The likelihood of success for the operation of five rockets on the mother plane and the launching of the research plane was considered to be very low.[10]

Creating the NB-52B

North American X-15 Rocket Plane

The North American X-15 was designed as a research vehicle for the investigation of (1) hypersonic aerodynamics, (2) structural characteristics under high aerodynamic heating conditions, (3) effectiveness of reaction type controls in regions where conventional aerodynamic controls are ineffective, and (4) psychological and physiological piloting problems associated with "weightlessness" conditions of space flight.[11]

When North American began to develop the X-15, it was assumed that the rocket plane would be air launched. Like the X-

X-15-1 equipped with a pair of Reaction Motors XLR-11 engines sits on Rogers Dry Lake in 1960. (Air Force Flight Test Center History Office)

2, the X-15 was designed with skid landing gear. The X-15s were built with a two-wheel nose gear and a pair of extendable skids at the rear. Later in the program, a third skid was added underneath the ventral fin.

One of the primary considerations in the choice of the launch airplanes was the weight of the X-15. The initial configuration of the X-15 weighed 14,500 pounds at launch for a glide flight and 34,400 pounds when loaded with rocket propellants.

The X-15 was designed to use the throttleable Reaction Motors XLR-99 rocket engine. The X-15 pilot could vary the thrust from 30,000 pounds up to 58,000 pounds. The rocket engine burned anhydrous ammonia and used liquid oxygen (LOX) as the oxidizer.

The development of the XLR-99 rocket engine took longer than had been anticipated. Due to delays in the development of the XLR-99, program managers decided to substitute a pair of Reaction Motors XLR-11 rockets to allow initial flight tests. These were the same engines that had been used in the X-1 program. They were refurbished for use in the first two X-15s. The eight chambers of the XLR-11 engines provided thrust in 2,000 pound increments, up to 16,000-pounds of thrust.

The first two X-15s were delivered with conventional air data booms on their noses. The booms carried small vanes that detected the attitude of the X-15 relative to the oncoming air. Later in the program the air data booms were replaced with a ball machined from solid Inconel-X for operation at high temperature. Sensors measured the ram air pressure in three holes on the front of the ball. Small motors rotated the ball to equalize the pressure in the three holes and determine the attitude of the X-15. The ball was referred to as the "Q-ball" or "hot nose".

Selection of the B-52 Stratofortress

At first, the mothership for the X-15 was expected to be the Convair RB-36. The X-15 would have been mounted in the bomb bay of the B-36. The performance of the B-36 would have allowed the 34,000-pound X-15 to be launched at an altitude of 35,000 feet and Mach-0.7.[12]

In 1957, The National Advisory Committee for Aeronautics (NACA) directed North American Aviation to change the mothership from the B-36 to the Boeing B-52 Stratofortress. The B-36 fleet was going to be phased out of service at about the time that the X-15 began flying. Spare parts would become difficult to acquire and the prospect of maintaining a single Peacemaker long after the type had been retired was daunting.

The X-15 could not be mounted in the bomb bay of the B-52. It would be mounted on a pylon under the wing of the Stratofortress between the fuselage and the inboard engine nacelle. This introduced a number of complications. The X-15 pilot would have to board the rocket plane before take-off. The B-52 flaps could not be used on take-off or landing. There were no windows on the B-52 that provided a view of the X-15 on the pylon.

The "flutter and noise engineers" were concerned that the noise from the B-52 engine pod might damage the X-15. Technicians mounted a dummy model of the X-15 on a B-52 wing and ran the engines for ten hours. Concrete ballast representing the X-15 was hung on a B-52 wing and dropped to see what effect it would have.

The improvement in performance conferred by the 72,000-pounds of thrust from the eight J57-19 jet engines of the B-52 raised the launch altitude to 45,000 feet and the speed up to Mach-0.9. It also allowed the weight of the X-15 to be increased.[13]

Two B-52s were assigned to the X-15 program. B-52A-1-BO, 52-0003 and RB-52B-10-BO, 52-0008 were sent to North American for modification in 1958.

RB-52B, 52-0008 was the eighth production Stratofortress. It was manufactured at Seattle. The RB-52B model was equipped

Proposed B-36 Carrier for X-15. (Author's artwork based on line drawing from AFFTC/HO)

RB-52B, 52-0008 at Boeing in 1955. (Boeing Archive Photo P14808)

to carry a pressurized crew module in the bomb bay for aerial photography. It was accepted by the Air Force on March 22, 1955, but its first flight was not flown until June 11. RB-52B, 52-0008 was assigned to Edwards Air Force Base in September 1955 for bomb/navigation systems testing.

In late 1955, the designation of 52-0008 was changed to JRB-52B and then to JB-52B. The new designation indicated that the bomber had been modified for flight test and that its reconnaissance capability was no longer supported.

JB-52B, 52-0008 tested the IBM MA-2 BRANE navigation system, which was the definitive Bomb/Navigation System for the B-52 fleet. Several pieces of equipment that were in development for use in later models of Stratofortress were installed in the JB-52B, including the AJA-1 True Heading Computer System, the AN/ASB-9 Bombing Navigation system, and the AN/APN-89 Radar System. The AN/ASB-9 and AN/APN-89 were used operationally in the B-52E and later models.

The Air Materiel Command received JB-52B, 52-0008 on July 3, 1958. Instrumentation recording equipment was removed. A spare VHF communication radio and a spare OMNI navigation radio were left aboard the airplane. The AJA-1 True Heading Computer System, the AN/ASB-9 Bombing Navigation System, and the AN/APN-89 Radar System were removed. Equipment related to operations of standard bombers was left inoperable on

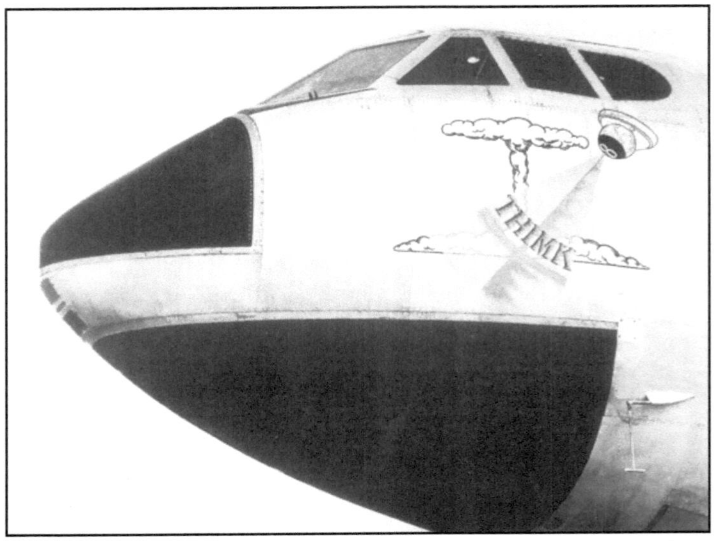

RB-52B, 52-0008 was used to test bombing and navigation systems that were put into service on later models of the Stratofortress. (Walter Boyne)

the JB-52B. The ALE-1 chaff dispenser was not installed. The wing and tail surface de-icing system was deactivated. The K-38 camera was not installed. The windshield wiper motor was disabled, and the wiper blades and arms were removed.[14]

52-0008 was transferred to the Aeronautical Systems Center on December 12, 1958.[15] It arrived at Plant 42 on December 13, 1958 and was moved into the North American hangar on January 6, 1959.[16] It had flown 110 times and accumulated 533 flight hours.

Modifications to the NB-52B

The NB-52B was modified as follows:

- A pylon weighing 1,170 pounds was added under the right wing, inboard of the engine nacelles and capable of supporting 50,000 pounds. The pylon was equipped with three shackles to support the X-15. Umbilical attachments supplied the X-15 with liquid oxygen, nitrogen gas for pressurizing the cockpit, breathing oxygen, and electrical power. A forty-gallon liquid nitrogen tank in the pylon produced nitrogen gas to pressurize the X-15 cockpit. Three release mechanisms insured that the X-15 could be released from the pylon. The primary release mechanism was hydraulic, powered by an accumulator mounted in the pylon. The secondary mechanism was pneumatic with its high-pressure nitrogen gas bottles located in the bomb bay of the NB-52B. A mechanical release was provided as backup. The design of the pylon placed the cockpit of the X-15 in front of the leading edge of the wing of the NB-52B so that the X-15 pilot could eject while it was still attached to the NB-52B.
- A 1,500-gallon liquid oxygen tank was installed in the bomb bay. It was divided into two sections, a climb tank and a cruise tank. The climb tank was sealed to prevent evaporation of the liquid oxygen. The cruise tank was vented

A pair of seventeen-inch Kintel television monitors provides the launch panel operator with views of the X-15 from the cameras mounted on the right side of the Stratofortress' fuselage. Above the monitors are gauges showing the cabin pressure of the X-15, the conditions of the pilots' suit, and the pressure of the hydraulic launch system. An altimeter is mounted at top center. A gauge presenting the XLR-99 engine pump lube oil temperature is mounted to its right. (NASA Photo E6845)

Launch panel operator's station on the NB-52B photographed about May 10, 1961. The seat originally used by the ECM systems operator is to the left. The upper left panel contains the push button controls for RADIO & NAVIGATION, LAUNCH OPERATOR, AN/APN-81 SYSTEM, LAUNCH SYSTEM, LOX TOP-OFF SYS, and X-15 SYSTEM, The center panel contains the controls for INST, MISC, X-15 HTR, TELEVISION, FCS, and AFCS. Some of the buttons have been physically disabled with clamps. The lower panel contains toggle switch controls for LAUNCH OPERATOR, TELEVISION, and X-15 SYSTEM. A number of the switches are equipped with safety covers. (NASA Photo E6840)

The launch panel operator faced an array of handles controlling the flow of liquid oxygen in the top-off system below the left monitor. (NASA Photo E6846)

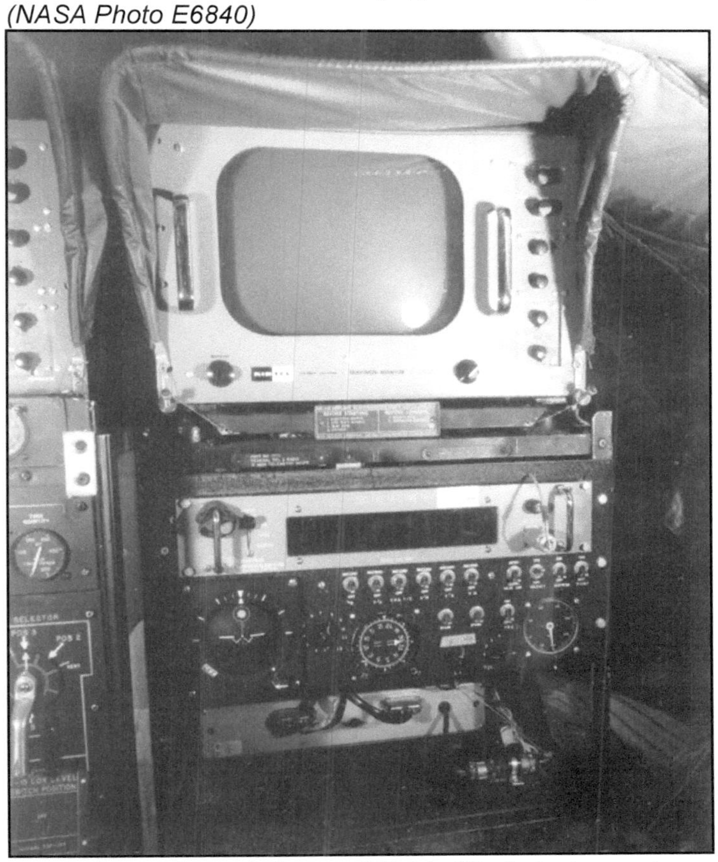

A set of recorder switches and flight attitude instruments were mounted on the panel below the right monitor. (NASA Photo E6855)

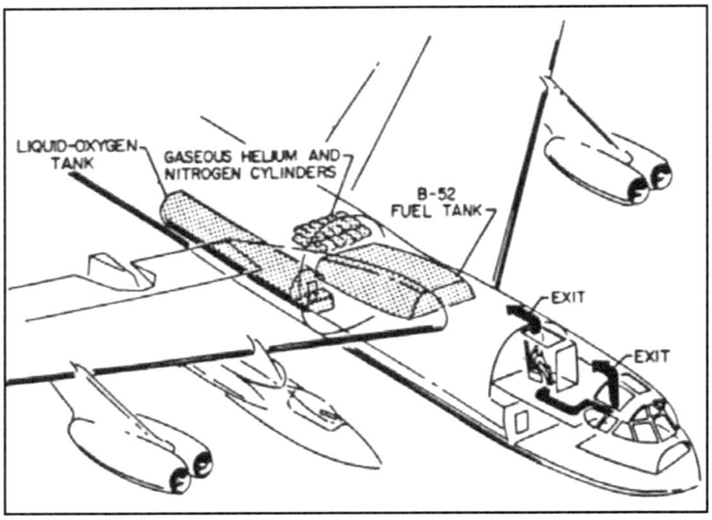

Illustration of modifications to the NB-52B from the X-15 Pilot Escape Manual (AFFTC/HO)

to the atmosphere to allow evaporation of the liquid oxygen in that tank, which cooled the liquid oxygen and increased its density. In the minutes before launch, the launch panel operator would switch from the climb tank to the cruise tank to maximize the amount of oxidizer in the X-15. The tanks were equipped with a jettison line and two vents that extended out the left side of the fuselage.

- Ten vents were installed on each side of the fuselage to prevent the accumulation of fumes in the bomb bay.
- Closed circuit television cameras, a film movie camera, and lights were installed on the starboard side of the fuselage. Two television cameras were installed, one forward of the X-15 and another aft of the wing. The movie camera was

The X-15 launch control panel is at the lower left corner of the pilot's control panel in the NB-52B. The primary launch switch is under the switch cover at the upper left corner of the panel. If the primary hydraulic launch system failed to launch the X-15, the B-52 pilot could activate the pneumatic alternate launch system by slipping the wire trap off the alternate launch handle and pulling the handle up. (NASA Photo E6842)

A North American X-15 Model was drop tested in the High Speed 7 x 10 foot Wind Tunnel at NASA Langley on February 15, 1958. (NASA Photo EL-1996-00114)

installed near the aft television camera.

- An AN/APN-81 navigation computing radar system was installed to provide input to the stable platform system. The stable platform established the position and direction of flight of the NB-52 and fed that information into the inertial navigation system of the X-15 in the moments before launch.
- A notch was cut into the trailing edge and inboard flap of the starboard wing to accommodate the vertical stabilizer of the X-15. The inboard flaps were disabled. The outboard flaps remained functional, but were not used in flight.
- 395 pounds of stiffeners, channels, ribs, rollers, and angles were added to reinforce the structure of the right wing.
- The No. 3 main fuel tank located in the wing above the pylon and the two cells of the mid-body fuel tank were removed.
- Eighteen nitrogen bottles and nine helium bottles were installed in place of the mid-body fuel tank. The helium gas was used to pressurize the liquid oxygen top-off system. The nitrogen gas was used to operate the valves of the top-off system.[17]
- A launch panel was installed on the upper deck of the NB-52B at the Electronic Counter-Measures operator's station. The launch panel fed altitude and velocity data to the X-15 inertial guidance computer until launch. A pair of television monitors allowed the launch panel operator to observe the liquid oxygen top-off overflow and check the function of the X-15 flight control surfaces.
- A Plexiglas astrodome was installed on the right side of the upper deck to provide the launch panel operator with a view of the X-15.
- A panel of five launch indicator lights was installed above the pilot's instrument panel to coordinate the pilots of the

NB-52B and X-15. Three lights confirmed that the pylon hooks had securely engaged the X-15. An amber "ready to launch" light and a green "launch" light were controlled by the pilot of the X-15. The "ready to launch" light indicated that the X-15 systems had been configured for launch. The "launch" light indicated that the X-15 pilot was ready to begin the final portion of the launch countdown.

- The right hand crew station on the lower deck was removed.
- The tail gun turret and tip tanks were removed.
- Wheels, tires and brakes designed for higher speeds were installed, raising the ground speed limit from 180 knots to 217 knots.[18]

The designation of 52-0008 was changed to NB-52B to indicate that it was so extensively modified that it could not be returned to stock configuration. The modified NB-52B returned from Palmdale to Edwards Air Force Base on June 8, 1959.

Those who worked on the NB-52B coined a number of nicknames for the flying launch pad; Zero Zero Eight, Double-Oh Eight, and Balls Eight among them.

Over the years, numerous systems aboard the NB-52B were replaced with systems from later models of the Stratofortress. Gauges in the fuel system are from the B-52C. The ejection seats are from the B-52D. The drag chute accepts the parachutes used in the B-52G, and the landing gear and brakes are the type used on the B-52H.[19]

NB-52A, 52-0003

Initial X-15 operations were conducted with Boeing NB-52A Stratofortress, 52-0003. The NB-52A first carried the X-15-1 aloft on a captive-carry flight on March 10, 1959. Three attempts to launch the X-15-1 on its first glide flight were aborted on April 1, April 10, and May 21. The X-15-1 was successfully launched on its first glide flight on June 8, 1959. The NB-52A carried the X-15-2 nine times and launched it on three powered flights while North American installed XLR-11 engines in the X-15-1 in the second half of 1959.

Boeing NB-52A Stratofortress, 52-0003 was the first mothership for the X-15 program. It first carried the X-15-1 aloft on a captive-carry flight on March 10, 1959. (Bob Dorr via Greg Spahr)

An explosion had occurred in the number one combustion chamber of the lower XLR-11 engine shortly after the X-15-2 was launched on its third flight on November 5. Scott Crossfield made an emergency landing on Rosamond Dry Lake. The X-15-2 landed heavy and its fuselage broke behind the cockpit. It was sent to North American in El Segundo to be repaired.

X-15 Operational Procedures

X-15 Preflight Inspection

Preparing the X-15 and NB-52B for a mission was a complex undertaking, requiring the coordinated efforts of dozens of technicians for many hours. Scott Crossfield described the early morning preparations for an X-15 launch in his book *Always Another Dawn*:

> The scene surrounding the preparations for an X-15 launch has been compared to a circus and to the packing up of a carnival. Work is begun early in the night and continues through morning. The area is bathed in light from the floodlights on the towers. The huge bare-aluminum airframe looms over the mechanics swarming around the B-52, preparing it for flight. The disciplined air of military operations is offset by the garish Day-Glo orange vertical tail, cockpit, and engine inlets of the B-52. Lights show in the cockpit windows as the aircraft ground crew goes through the lengthy pre-flight checklist.
>
> Beneath and around the bomber there are not less than twenty-five trucks and carts, and probably a hundred men. Some of them are concentrating their efforts on the B-52, but most of them are clustered around the black missile hanging from the pylon under the wing. Vapors from the super cold liquid oxygen being fed into the X-15 swirl around the working crew.
>
> By the time the X-15 pilot is suited up and prepared to enter the cockpit of the black rocket, the ground crew wraps up the preparation of the vehicles. The ground carts are pulled to one side. Cables and hoses are pulled from the connections on the X-15. The access panels are all replaced on the X-15 and its pylon.[20]

X-15 Pylon Preflight Inspection

During the day prior to an X-15 flight, an extensive preflight inspection was conducted of the rocket plane and its mothership.[21] Access panels on the pylon and the inboard right wing were removed. The pylon was checked for damage, loose or missing rivets, and corrosion. The pylon to wing seal and attaching bolts were checked for security.

The X-15-1 sits under the pylon of the NB-52B before mating in November 1959. (NASA photo E5084)

Technicians inspected the many components of the pylon release mechanism; the push-pull rods, bell cranks, cables, pulleys, sectors, shackles, and sway brace assemblies. Pylon hydraulic and pneumatic lines, selector valves, actuating cylinders, hydraulic accumulators and storage cylinders were inspected for proper mounting, security of attachment, leakage and damage.

The central component of the primary release mechanism was the pylon hydraulic accumulator. It was charged to a pressure of 1,800 pounds per square inch.

A backup pneumatic release system was provided in the event of failure of the hydraulic release mechanism. The pneumatic system's nitrogen cylinders were located in the bomb bay. They were charged with high-pressure nitrogen gas.

Pylon electrical switches, terminal boards, wire bundles, and connectors were examined for cleanliness, chafing, damage, adequate clearance, and security.

Launch Panel Operator's Station

The Launch Panel Operator's observation window and astrodome were inspected for cracks, cleanliness, and security of mounting. The control panel handles and switches were checked for freedom of operation.

Technicians checked the legibility of the markings on the instrument and control panels. The instrument cover glasses, indicator hands, range markings, and mountings were inspected for damage and security. Damage and security checks were made of the instrument panel wire bundles, connectors, terminal boards, and chafing strips. Instrument transmitters were similarly checked.

The X-15 pilot's breathing oxygen system cut-off valve on the aft bulkhead of the launch panel operator's compartment was wired in the "on" position to prevent any accidental interruption.

The markings on the electrical control and fuse panels were checked for legibility. The switches, fuses, circuit breakers, and electrical controls were inspected for damage and to see that they had been properly positioned for the upcoming flight. The indicator lights were checked for proper installation. The supply of spare lamps and fuse holders was restocked.

The closed circuit television mounts, interconnecting cables, connectors, terminal strips, and junction boxes were inspected for the presence of foreign material, damage, corrosion, and security.

X-15 Related Systems on the NB-52B

Several X-15 related systems aboard the NB-52B required extensive inspection before each flight. The windows on the television and movie camera fairings were cleaned and checked for cracks and the fairings were examined for proper attachment. The camera fairing, window, and astrodome defrosting system ducts, valves, thermostats, mounting brackets, and clamps were checked for damage and security.

The television cameras were equipped with blue filters that were checked for proper installation and whether they had faded to the point of needing replacement. The flood lamps on the right side of the fuselage were checked for damage and security.

The liquid oxygen tank in the bomb bay was checked for damage and any evidence of leakage. Associated liquid oxygen lines, control valves, regulators, relief valves, clamps, and fittings were inspected for damage and security of attachment. The filler

The X-15-1 was mated to the pylon of the NB-52B for the first time in November 1959. (NASA photo E5087)

receptacle and pylon breakaway fittings were also checked. The breakaway fittings allowed for the separation of the X-15 from the top-off system at the moment of launch. The tanks, lines, valves, regulators, relief valves, filler receptacle, clamps and fittings of the helium pressurization system and the nitrogen gas control system were inspected.

Electrical wiring, terminal boards, and connectors from the X-15 pylon to the NB-52B were checked for cleanliness, chafing, damage, adequate clearance, and security.

All auxiliary UHF radio, AN/AIC-10 interphone, telemetry system, and X-15 stable platform system components, mounts, control knobs, dials, interconnecting cables, antenna, connectors, and terminal strips received an inspection for cleanliness, damage and security. Two UHF antennae and the telemetry blade antenna were checked for damage and security.

The AN/APN-81 Doppler radar system components, mounts, control knobs, dials, interconnecting cables, antenna, connectors, and terminal strips would receive an inspection for cleanliness, damage and security. The Doppler radar system antenna, wave guide, coupler, receiver, and transmitter were pressure tested. The system was sealed at the atmospheric pressure of the flightline by opening and closing the air bleed valve in the bombardiers entrance hatch before flight.

The forward, center, and aft instrumentation cameras in the NB-52B fuselage and the instrumentation camera in the aft fairing of the X-15 pylon were checked for damage and security.

X-15

Cockpit windshield anti-fogging and space suit ventilation system tanks, mounting brackets, relief valves, transmitters, indicators, regulators, quick disconnect fitting, and associated lines were checked for damage and security of mounting. The systems were then pressurized to 2,200 pounds per square inch with nitrogen gas.

Technicians checked the Auxiliary Power Unit (APU) feed compartment heat control system lines, valves, brackets and clamps for damage and security of attachment.

The X-15 pilot's breathing oxygen system lines, clamps, and mounting brackets were examined. The breathing oxygen was equipped with a breakaway fitting was also inspected for damage, chafing, and security of attachment.

The combustion chamber of the XLR-99 engine was cooled with nitrogen gas. The cooling system nitrogen tanks, lines, valves, check valves, mounting brackets, and clamps were inspected.

Preparations for Powered Flight

As the "Hot Papa" serviced the X-15 with its propellants, hoses directed sprays of water onto the concrete behind the wing of the NB-52B. Any overflows of hydrogen peroxide, water alcohol, and liquid oxygen were directed into the water to dilute the spills and prevent any fires from starting. A retaining wall surrounded the area to contain any large spills. Fitzhugh Fulton described the scene, "A lot of times there would be leaking ammonia fumes. There would be a terrible smell around the airplane."

Power On Check

Following engine start, a power on check was made of the NB-52B and X-15 systems. The hydraulic, pneumatic, and mechanical release mechanisms of the X-15 pylon were given an operational check. Proper operation of the X-15 auxiliary power unit feed compartment heat control system was verified. Other systems requiring operational checks included the NB-52B cockpit windshield anti-fogging nitrogen regulator heating system, the liquid oxygen system, NB-52B fuel system, space suit nitrogen ventilation system, X-15 cockpit windshield anti-fogging system, X-15 pilot breathing oxygen system, and the television camera and observation window defrost system.

The operation of the instrumentation cameras in the NB-52B fuselage and the X-15 pylon and the forward and aft drop cameras was checked. The operation of the heating blankets installed in the forward and aft drop camera fairings was confirmed. An operational check was performed of the forward pylon release hook heating system.

A "Push for Light Test" confirmed the function of all warning and indicator lights. The externally mounted television floodlights were turned on.

Instruments aboard the NB-52B were checked to insure that they indicated the proper temperature, pressure, and quantity values. A leak check was made of the airspeed, altimeter, and Mach-meters at the pilot, co-pilot, and launch panel operator stations.

Operational checks were made of the X-15 stable platform external power supply, the closed-circuit television system, the auxiliary UHF radio, AIC-10 interphone, and telemetry system. The stable platform wiring was subjected to a high voltage test.

An operational check was made of the AN/APN-81 Doppler radar system. The AN/APN-81 radar produced considerable heat, so the equipment racks were checked for adequate cooling.

The X-15-1 was mated to the pylon of the NB-52B for the first time in late 1959. (AFFTC/HO)

Immediately Prior to Flight

As each crew member boarded the NB-52B, he installed the M-3 ejection seat initiator linkage and extracted the pin of the red "Remove Before Flight" streamer, arming his ejection seat.

B-52 Flight Operations

Start and Taxi

Twenty-two pieces of rotating equipment had to be started in order to prepare the NB-52B for flight; eight jet engines, ten air-driven hydraulic packs, and four alternators. That is more than any airplane built before or since.

Normal B-52 Flight Manual procedures were used except for the engine start sequence. To minimize the exposure of the X-15 to the engine noise of the inboard engines (engines 5 and 6) on the right side of the NB-52B, the four engines on the left side were started first, inboard to outboard. Then the engines on the right side of the airplane were started, also inboard to outboard.

The NB-52B listed slightly toward the X-15 while taxiing.

Take-off

To compensate for the weight and drag of the X-15, trim settings were adjusted to left wing down and slight right rudder for take-off. The wings were leveled when the ailerons became effective as the airspeed increased.

Take-offs were made using water injection[1] only on the outboard four engines of the NB-52B, or on all but engines 5 and 6 to reduce the volume of the noise that the X-15 was subjected to. The acoustic energy of the noise from the engines operating with water injection was so intense that there was concern that the X-15 might suffer structural damage if it were used on the adjacent engines.

If the calculated ground roll for the Stratofortress exceeded 12,000 feet, lakebed runway 17-35 was used for take-off. Water injection was necessary for take-off from the concrete runway when the ambient temperature exceeded 90 degrees F. The lakebed runway was used for take-off by NB-52A, 52-0003 when the air temperature at Edwards Air Force Base exceeded 90 degrees F. because it did not have water injection.

Gross weight of the NB-52B at take-off was about 305,000 pounds. The basic airplane weighed about 175,000 pounds. 128 gallons of oil added 960 pounds. The crew was estimated to weigh 920 pounds. The fuel tanks contained about 80,000 pounds of fuel. A fully fueled X-15 weighed over 34,000 pounds and the bomb-

1 Most B-52s were equipped with water injection for the engines. The conversion of water to steam absorbs heat and reduces the temperature in the combustion chamber. The expanding steam adds significantly to the thrust of the engine. Water injection also substantially increases the level of noise created by the engines.

bay tank contained almost 15,000 pounds of liquid oxygen.

The main runway at Edwards Air Force Base is aligned on a heading of 40 degrees magnetic. The standard runway for X-15 operations was Runway 04, which has an overrun onto Rogers Dry Lake. Runway 22 was not normally used.

The flaps on the NB-52B were disabled, so it took off a lot faster than any other B-52. The normal no flap take-off speed was 181 knots but the maximum speed allowed with the gear extended was 200 knots. After take-off it was important to retract the landing gear immediately as the NB-52B accelerated to avoid damaging the landing gear doors during the retraction sequence.

Climb

Climb was established at 260 knots until reaching 0.78 indicated Mach number, which was maintained at higher altitude. The NB-52B made a climbing turn to the left to make it possible for the pilot of the X-15 land on the lakebed in the event of an emergency launch. The Stratofortress remained near Rogers Dry Lakebed until it reached an altitude of 38,000 feet. The chase airplanes joined up with the NB-52B during the ascent to launch altitude.

Early X-15 flights were launched near Edwards Air Force Base. On those flights, the Stratofortress flew in a holding pattern at 38,000 feet until seventeen minutes before launch. Then the NB-52B climbed to the launch altitude of 45,000 feet. The climb was conducted at the highest possible speed to allow the Lockheed F-104 Starfighter and North American F-100 Super Saber chase airplanes to stay in position. The top speed of the Stratofortress was only a little higher than the stall speed of the chase planes.

The indicated airspeed[2] during the last two thousand feet of climb dropped as low as 205 knots. After leveling off at 45,000 feet, the NB-52B accelerated to 0.82 indicated Mach number.

The modified NB-52B suffered a seven-percent reduction in miles per gallon compared to a standard B-52. Carrying the X-15, the NB-52B got sixteen percent less mileage. Despite the additional drag and the removal of the number 3 main wing tank and the mid-body tank, the available fuel easily accommodated a typical launch flight.

The NB-52B encountered Mach buffet[3] at a lower airspeed than standard B-52s. At typical flight weight and launch altitude,

The X-15-1 was mated to the pylon of the NB-52B for the first time in November 1959. (NASA photo E5087)

Mach buffet was encountered at 0.82 indicated Mach number. The NB-52B had a tendency to roll off to the right at Mach-0.85, which was easily corrected by the application of left aileron.

X-15 Checkout Prior to Launch

Shortly before launching the X-15, the pilots of the NB-52B rolled it left and right to bank angles of forty degrees so that the X-15 pilot could monitor the response of the stability augmentation system to the rapid change in attitude.

The X-15 pilot fed manual inputs into its reaction control system to establish its proper function before launch. At indicated airspeeds above 240 knots some of the exhaust from the reaction control system was sucked into the NB-52B aft alternator cooling air duct on the right wing leading edge and into the air ducts on the right side of the fuselage of the NB-52B. The exhaust of the hydrogen peroxide fueled thrusters was just steam and its ingestion into those openings was not a problem.

The X-15 pilot cycled its flight control surfaces before launch. Their operation did not noticeably affect the NB-52B. The extension of the X-15 speed brakes generated a slight airframe buffet. Tests of X-15 flap extension produced a small nose-down trim change. Initially, the extension of the nose gear produced a "thump" that could be felt and heard inside the NB-52B. The gear extension sequence was later modified to eliminate the thump.

2 Indicated airspeed is the value presented to the pilot on his airspeed indicator. The airspeed indicator measures the pressure of air entering the front of the pitot tube. Because the density of air changes with altitude and temperature, the airspeed indicator shows a speed that differs from the true airspeed of the aircraft. This can be corrected for to determine the true airspeed, but for flight operations, the indicated airspeed has more relevance to lift and control characteristics.

3 Mach buffet is a shaking of an airframe resulting from the interaction of shock waves forming on parts of the airframe as it approaches the speed of sound. The X-15 pylon mounted between the fuselage and engines caused this effect to become noticeable at lower airspeeds than was typical of a standard B-52.

Launching the X-15

When the X-15 was launched, the right wing of the NB-52B was suddenly relieved of 34,400 pounds. The weight of the X-15 was offset forward and to the right of the center of gravity of the Stratofortress, so launching the X-15 produced a roll to the left and a pitch up. The application of half aileron to the right and forty pounds of push force on the stick would counteract the attitude change.

Fitzhugh Fulton described the NB-52B pilot's perspective of launching the X-15:

> It was a great feeling when you dropped the test vehicle, a great, big thump, just like dropping a bomb, and then you'd listen for the rocket engine to fire up. Some of the X-15 pilots were quicker than others. You'd hear the rocket engine fire up, and then just a few seconds the rocket airplane would come out in front the nose.
>
> We had the best view of the whole operation, to see the rocket plume coming out there and the airplane accelerating away at some fast speed.
>
> We'd listen on the radio and see if there was any problem. In some cases there were a few runs where they had to land on the lakebed someplace like near Las Vegas or near Tonopah, Nevada. If that happened, we'd circle down behind the chase airplanes and take a look ourselves.

Post-Launch Operation

Not all of the liquid oxygen in the 1,500-gallon tank in the bomb bay of the NB-52B was required to replenish the X-15 before its launch. The remainder was jettisoned before the NB-52B landed. The jettison would take five to fifteen minutes. The landing gear was kept retracted until the jettison is complete to prevent liquid oxygen from flowing into the open left-rear main landing gear well.

The descent was made at 0.75 indicated Mach number until 250 knots indicated airspeed was reached. The descent then proceeded at 250 knots indicated airspeed. Airbrake position two was normally used for descent.

Fitzhugh Fulton related that, "If it was a routine launch, then the X-15 or lifting body would be on the ground and all the ground crew around it before we could get back to Edwards. We'd come back and make a flyby of the lakebed and then go ahead and land."

If the X-15 was not launched, the X-15 pilot jettisoned the liquid oxygen, water-alcohol fuel, and hydrogen peroxide. The hydrogen peroxide jettison was delayed until the Stratofortress entered the landing pattern, because it powered the auxiliary power unit that was needed to operate the X-15 control surfaces in the event of an emergency launch.[22]

Landing the NB-52B

Landings with the X-15 attached were performed according to the no-flap procedure described in the B-52 Flight Manual. Flare speed was based on the gross weight of the airplane. The downwind leg of the landing pattern was flown 30 knots faster than the flare speed. The base leg was flown 20 knots faster than flare and the final approach was started at 10 to 15 knots above flare. Airspeed was allowed to bleed off in the final approach so that the flare speed was achieved over the threshold of the runway.

The landing attitude without flaps was nose high compared to a standard Stratofortress. Most of the elevator control authority was used to hold the nose up on approach, so little was left to control the sink rate of the nose after the rear main gear contacted the runway. The nose settled rapidly and the nose gear smacked down hard on the runway. The plot extended the airbrakes after touchdown and deployed the drag chute when the NB-52B had slowed to 150 knots.

Post Flight

X-15 Pylon

Following flight the X-15 pylon was examined for loose or missing rivets and fasteners, cracks, and buckled skin that might indicate internal structural failure. The hooks, sway brace fittings, and quick disconnect fittings were checked for cleanliness, evidence of wear or deterioration. Protective covers were installed on the quick disconnect fittings. The pylon and surrounding areas of the NB-52B were checked for the presence of corrosion or corrosive liquids.

Launch Panel Operator's Station

The red "Remove Before Flight" streamer pin was inserted in the ejection seat immediately to prevent accidental firing the ground. The launch panel operator's compartment and observation windows were checked for cleanliness. The equipment racks, control and instrument panels, and consoles were checked for damage and the legibility of their decals and markings.

The launch panel operator's seat and escape hatch were inspected for damage and the presence of foreign material. The lanyard was disconnected from the M-3 ejection seat initiator.

X-15 Related Systems on the NB-52B

Liquid oxygen, helium, and nitrogen tanks, lines, mountings, valves, and fittings aboard the NB-52B were checked for damage, leakage, and security of attachment. All selector valves and control switches were placed in the appropriate positions.

All electrical switches and circuit breakers were placed in the proper positions. Spare lamp and fuse holders were re-stocked. All electrical equipment was examined for signs of overheating and security of mounting.

Shock mounted equipment, which had a limited range of movement, was checked for adequate clearance from surrounding structures. Connectors to the shock mounted equipment were

Hydraulic lifts were installed in the X-15 servicing area to mate the X-15 to the pylon on the wing of the NB-52B. The lifts are being raised up to the skids of X-15-1 on December 18, 1959. (NASA Photo E5104)

checked for security of attachment and proper setting of the safety interlocks.

Wire bundles were inspected for cleanliness, security, and proper routing. Terminal strips and junction boxes were checked for foreign material and loose terminals.

Instrument cover glasses, indicator hands, range markings, and mountings were examined for damage and security.

The television cameras and monitors were turned off and camera window covers were installed.

Communications, navigation, and telemetry equipment components were examined for evidence of overheating and security of mounting. Control knobs, switches, and dials were checked for proper position and security.

Periodic Maintenance of X-15 Related Systems

The pylon to wing attachment bolts were torqued to 9,000 to 10,000 inch pounds following every tenth flight.

Liquid oxygen flange bolts were checked for proper torque and liquid oxygen line welds and bellows were checked for cracks and leakage each twenty-five flight hours. Differential pressure tests of the nitrogen gas system filters were performed each fifty hours. The liquid oxygen lines were evacuated to 100 microns every six months. The liquid oxygen tank was vacuum checked every six months.

North American X-15

In early November 1959, the X-15-1, 56-6670 was mated to the NB-52B for the first time. Technicians conducted tests of the X-15 LOX tank liquid level sensing probe and NB-52B top-off control amplifiers of the Acoutica Associates automatic top-off liquid level control system. Automatic top-off cycles were accomplished from both the climb and cruise tanks in the NB-52B.

The NB-52B was assigned a ground crew from the 6517th Flightline Maintenance Squadron (Bomber). Capt. Maynard N. Heth commanded the 6517th. MSgt. Kenneth L. LaPean was the Flight Chief for the two NB-52 motherships. TSgt. Lester

Schroyer, TSgt. Harweda Terrill, SSgt. Lawrence Maharrey, Airman First Class Donald Kemmer, Airman First Class Vernon Barr, Airman Second Class Norman Smith, and civilians Robert Swanson, William Wyant, and Albert "Pappy" Rice served in the ground crew.

Camera Installation in NB-52B

The aft vertical camera on the NB-52A captured important movies of the engine explosion of the X-15-2, 56-6671 on its third powered flight. As a result, priority was placed on the installation of the aft vertical camera on the NB-52B as it was prepared for its first X-15 mission. Colonel Allen and Lt. Colonel Harold Russell drafted a directive that the camera installation be performed on a non-interference basis with the X-15 program. The camera installation was considered a safety of flight item. [23]

The installation was budgeted for 350 man-hours and $1,500 in materials. Lt. Colonel K. C. Dudley reported to Richard Harer, the project engineer for the X-15, that attempts to determine whether North American had a modification kit were thwarted by North American's pre-occupation with making repairs to the X-15-2, so the estimate assumed that they did not have the kit. Work on the camera installation was scheduled to begin on November 24, 1959 and to be finished on December 10 in time for the first launch from the NB-52B.[24]

XLR-11 Powered X-15 Flights

As 1959 drew to a close, the X-15 program entered into the initial envelope expansion phase. The prime objective of the envelope expansion phase was to attain the maximum speed and altitude capabilities of the XLR-11 powered X-15 in the most rapid manner consistent with flight safety. The full potential of the X-15 could not be achieved with the interim XLR-11 rocket engine installation.

Seasonal rains closed the lakebeds until mid-December. The NB-52B first carried an X-15 in flight on December 16, 1959. It was an attempt to launch Scott Crossfield on the second flight of the X-15-1, 56-6670, which would be its first rocket-powered flight. Captain Charles Kuyk piloted the NB-52B and Major Jack Allavie was the co-pilot. Take-off was at 8:30 in the morning. After take-off, the radio in the X-15 failed, reducing communication to the clicks of the microphone key, one for yes, two for no.

When Crossfield pressurized the propellant tanks, the LOX tank pressure would not stabilize. The helium regulator was malfunctioning. Initially, the pressure stayed within limits for launch, but shortly before the drop, the pressure fell below the limits for safe operation of the XLR-11 rocket engines. Q. C. Harvey, in the X-15 operations room on the ground, queried Crossfield whether they should proceed with the launch. Crossfield replied with a pair of mike clicks. Harvey asked for confirmation that Crossfield did not want to launch. Crossfield's response was a single click, and the mission was aborted five seconds before the drop. Another attempt to launch the X-15-1 on December 21 was canceled.[25]

The next launch attempt was over a month later, on January 23, 1960. Telemetry difficulties delayed the take-off. Then an airplane crash closed the main runway.

Major Fitzhugh Fulton and Captain Kuyk launched Scott Crossfield at an altitude of 45,000 feet over Rosamond Dry Lake. It was the second time that Fulton piloted an X-15 launch mission. He would eventually pilot ninety-four X-15 launches.

The second flight of the X-15-1 was the first fully successful rocket-powered X-15 flight. The XLR-11 rockets propelled it to Mach-2.53 (1,669 miles per hour) and 66,844 feet altitude. The only pilots who had flown faster than Crossfield did on this flight were Pete Everest and Mel Apt in the X-2.[26]

North American repaired and strengthened the fuselage of the X-15-2. It returned to Edwards Air Force Base in January 1960.

The NB-52B carried the X-15-2 aloft for a powered flight attempt on February 4. There were numerous systems problems on the X-15. The cockpit would not pressurize. The radio failed. The bearings of one of the APUs seized. The pressure of the water alcohol fuel tank sagged, and the jettison failed to completely empty the tanks of the X-15. The mission was aborted, and the dead weight on the pylon was much greater than usual when the NB-52B landed.[27]

Stable Platform Instrumentation Installation

For flights out of the atmosphere, ordinary flight attitude instruments are useless. The X-15 required an inertial navigation system to track its attitude. The stable platform system used data from the instruments of the NB-52B to establish a fixed frame of reference for the inertial navigation system of the X-15. It was a challenging task to get all of the instrumentation working together to fix the frame of reference. The process was referred to as erecting the stable table. Once the stable platform was erect, it fed position and velocity data into the inertial navigation system of the X-15 until launch. Colloquially, the stable platform was also called the stable table.

NASA supplied the instrumentation for the stable platform system of the X-15. The instrumentation was installed in the NB-52B in February 1960. An XG recorder was mounted on the forward bulkhead of the engineer's compartment. Other equipment installed included a nine-channel Mod Box, a calibration and zero control box, and eighteen channel w.s. control box, a NASA timer, a twenty-six channel oscilloscope, an instrument mounting plate, and a film magazine. The cost of the instrumentation was $26,550. 1,200 man-hours were budgeted for the installation. [28]

An additional wiring harness for the stable platform was installed after the nineteenth flight of the X-15-1 was flown on December 9, 1960. The NB-52B was off flight status for ten days as 160 man-hours were expended on the installation. The cost of

The X-15-1 is lowered to the ground on December 18k, 1959. (NASA Photo E5110)

The rear hydraulic lift reaches the ground first as the X-15-1 is lowered to the ground in a level attitude on December 18, 1959. `(NASA Photo E5111)

the installation was $1,110. [29]

LOX Top-Off System Automation

As the NB-52B hauled the X-15 to the launch point, the liquid oxygen in the tank of the X-15 was evaporating. The X-15 liquid oxygen supply was replenished from the tank in the bomb bay of the NB-52B. North American installed instrumentation to automate the operation of the liquid oxygen top-off system in February 1960. Forty-eight man-hours were expended over three days on the installation. A "black box" was mounted on a bulkhead in the bomb bay and its associated wiring was routed through the pressure bulkhead into the lower crew compartment. [30]

X-15-2 Returns to Flight

The NB-52B carried the X-15-2 aloft again on February 11. Captain Jack Allavie was at the controls of the mothership, and Major Fulton was the co-pilot. The Stratofortress loitered at the launch altitude for the length of time that it would take to carry the X-15 to the launch sites used for future high performance flights. Despite the cold-soaking, the X-15 systems functioned properly, and the launch was conducted successfully. Although standard practice had been for the pilot of the Stratofortress to trigger the launch of the X-15, Crossfield triggered the launch himself. The procedure later became the standard method. The XLR-11 engines propelled the X-15 to an altitude of 88,116 feet. Crossfield leveled out and accelerated to Mach-2.22 (1,466 miles per hour) on a flight that was a complete success.[31]

Six days later, Fulton was at the controls of the NB-52B, and Allavie was the co-pilot for the launch of the fifth flight of the X-15-2. The upper XLR-11 engine of the X-15-2 shut down part way through the burn. The thrust available to the X-15-2 was cut in half, but the time of the burn was correspondingly lengthened. The reduced thrust resulted in a modest speed of Mach-1.57 (1,036 miles per hour), one of the slowest powered flights of the X-15 program. The X-15-2 climbed only a mile and a half above the launch altitude, reaching 52,640 feet.

High winds at Edwards Air Force Base canceled the next two launch attempts scheduled on February 25 and 26. It was not until March 17 that Major Allavie and Captain Kuyk launched Crossfield in the X-15-2 from the NB-52B again. The systems of the X-15-2 performed flawlessly on a flight that reached Mach-2.15 (1,419 miles per hour).

The ground crew turned the X-15-2 around overnight for another launch attempt the following day. Major Allavie and Captain Kuyk were assigned to fly the NB-52B. Just seconds before the drop, chase pilot Al White noticed alcohol leaking out of a drain from the engine bay. A fitting in the fuel plumbing had cracked and alcohol was pooling in the bay. There was great danger of an explosion of the vapors in the engine bay, so the launch was aborted.[32]

Joe Walker in the cockpit of an X-15 following a flight. (North American Rockwell)

Joe Walker Flies the X-15

North American delivered the X-15-1 to NASA in January 1960. NASA pilot Joe Walker would be the first pilot other than Scott Crossfield to fly the black missile. Walker had acquired plentiful rocket plane experience in the X-1E. To prepare him for his first flight in the X-15, he performed a ground engine run and a checklist "dry run". The entire flight crew participated in the dry runs to familiarize each new X-15 pilot with the procedures used during the launch. The X-15 pilot, the NB-52B pilot and the launch panel operator conducted a simulation of the entire sequence of operations, including a short run of the auxiliary power unit of the X-15. [33]

The first NASA X-15 flight was scheduled for March 23, but bad weather forced the flight to be postponed for two days. On March 25, Major Allavie and Fitzhugh Fulton piloted the NB-52B around a racetrack course as Walker familiarized himself with the systems of the rocket plane before the launch. Walker flew the X-15-1 to Mach-2 (1320 miles per hour) and an altitude of 48,630, only slightly higher than the launch altitude.

Walker's flight marked the beginning of NASA's expansion of the envelope of X-15 operations with the interim XLR-11 engine installation. Four local flights were made in March, April, and May of 1960 to familiarize Joe Walker and Robert White with the behavior of the X-15. Each flight followed a triangular path, northeast toward the gold mining town of Randsburg, west toward the Tehachapi Mountains, and then southeast back to Rogers Dry Lake.

Four days after Joe Walker's first flight in the X-15, Scott Crossfield was launched from the NB-52B in the X-15-2. Fulton and Allavie piloted the Stratofortress. The flight investigated the behavior of the black rocket at the high angle of attack that would be encountered on pullouts from high altitude. Crossfield performed a series of dives and pullouts, pulling up to six gs[4] on

4 One g equals the force of gravity at the surface of the

The NB-52B made its first public appearance at the May 17, 1960 Edwards AFB open house with the X-15-1 mounted under its wing. The X-15-2, 56-6671 was displayed on the left side of the NB-52B. (Richard Lockett)

the pullouts.

The ground crew prepared the X-15-2 for its next flight in just two days. Crossfield made the seventh flight of the X-15-2 on March 31. Allavie and Fulton piloted the NB-52B.[34]

Launch Lakes

As the NASA pilots expanded the X-15 flight envelope, launches moved away from Edwards Air Force Base to dry lakebeds located to the northeast. Silver Lake was used for X-15 flights of about 100 nautical miles. Hidden Hills Lake was used for flights of about 130 nautical miles. Mud Lake was used for flights of about 200 nautical miles. Other dry lakebeds between Edwards Air Force Base and the launch lakes served as emergency landing strips.

Doppler radar Roll Limit Switches

The usual procedure on a launch flight was to fly from Edwards to the vicinity of the launch lake and then make a 180-degree turn to the launch heading. The pilot of the NB-52B placed it in a twenty-five degree bank to the left for five minutes.

On the initial flights to Silver Lake, a problem arose with the operation of the stable platform. The APN-81 Doppler radar system on the NB-52B measured the ground speed and course of the airplane to update the stable platform. The stable platform provided a frame of reference for the instruments of the X-15's inertial navigation system before launch.

The Doppler radar system was only accurate when the Stratofortress was flying straight and level. To prevent the radar from generating inaccurate speed data, it was equipped with roll limit switches that shut off the radar input to the stable platform at a bank angle of thirteen degrees.

During the turn to the launch heading, the stable platform was deprived of the critical input from the Doppler radar. After the NB-52B returned to straight and level flight, the Doppler radar took some time to resume proper function. This was normal operation for Doppler radar units on standard B-52 bombers. The radar took too long to return to normal operation after the Stratofortress assumed the launch heading. Failure to maintain erection of the stable table in this critical period sometimes resulted in an abort of the launch attempt.

In April 1960, Richard Harer, the X-15 project officer, directed that the roll limit switches on both NB-52s should be disconnected. The cost of the disconnection was budgeted for $16.00 and four man-hours.[35]

The NB-52B was placed on public display for the first time at the Edwards Air Force Base Open House on May 17, 1960. The X-15-1 was suspended from the launch pylon and the X-15-2 sat on the ground to the left of the Stratofortress. The NB-52A made a fly-by and then taxied past the crowd.

Last X-15-2 Flight with XLR-11 rockets

The NB-52B launched the X-15-2 six times with the interim XLR-11 rocket engines before it was sent back to North American for installation of its XLR-99 engine. It launched the X-15-2 on its last flight before conversion on May 26, 1960. It was also the first free flight test of the operation of the X-15's reaction control thrusters. The thrusters were powered by concentrated hydrogen peroxide that was converted to steam by passing it through a silver screen. The silver catalyst caused the hydrogen peroxide to break down into water and oxygen at high temperature.

Captain Charles Bock, Jr. and Major Allavie piloted the Stratofortress. Scott Crossfield flew the X-15-2. The flight was only intended to demonstrate that the reaction control thrusters functioned properly. The maximum altitude attained on the flight was only 51,282 feet, so the thrusters had no appreciable effect on the attitude of the X-15.

Navy Commander Forrest Petersen made his first X-15 flight on September 23, 1960. It was the thirteenth flight of the X-15-1. Jack Allavie and Fitzhugh Fulton piloted the NB-52B.

Earth

The X-15-2 was displayed on the left side of the NB-52B at the 1960 Edwards Air Force Base Open House. (W. Duncan via Greg Spahr)

Commander Petersen was launched from the Stratofortress near Palmdale, southwest of Edwards Air Force Base on September 23. Petersen pointed the X-15-1 to the northeast, toward the town of Boron. His flight plan called for him to make a left turn near Boron toward the town of Mojave and then another left turn to return to Edwards Air Force Base.

The XLR-11 engines shut down when Petersen had reached only Mach-1.68 (1,108 miles per hour). He was turning left over Boron when he lost power. The flight path for his familiarization flight had been selected to keep him positioned for a return to the lakebed in the event of just such an emergency. He continued the turn to reach the high key position for runway 18 on Roger Dry Lake with Joe Walker in the chase plane tucked in close beside the X-15.

Petersen reached the high key position at just 25,000 feet altitude, which was lower than the desired altitude. Walker urged him to tighten up his turn as they made a left 360-degree turn. Petersen arrived on final approach a little high on energy. He popped the speed brakes to set the X-15 down at the correct position on the runway. The Dryden Diary entry for this flight says, "Exciting but good".

Following its ninth flight, the X-15-2 was trucked back to North American in El Segundo for the installation of its XLR-99 rocket engine in September 1960. It returned to Edwards Air Force Base later that month.

The X-15-1 made two flights in September 1960 and two flights in October 1960. Major Robert White was scheduled for launch on his fifth X-15 flight on September 2, but the flight was

The NB-52B takes off from Runway 04 with the X-15-1 in the fall of 1960. The Day-Glo orange has been stripped from the vertical stabilizer and a six-foot wide Day-Glo orange band has been added to the rear fuselage. (AFFTC/HO)

Fitzhugh Fulton piloted the Stratofortress. Commander Forrest Petersen occupied the cockpit of the X-15-2. (U.S. Air Force Photo via Terry Panopolis)

X-15 pilots in 1960; Air Force Captain Bob Rushworth, NASA pilot John B. "Jack" McKay, Navy Commander Forrest Petersen, NASA pilot Joe Walker, NASA pilot Neil Armstrong, and Air Force Major Bob White. (North American Rockwell)

cancelled by bad weather before the NB-52B got off the ground. Problems with the Stability Augmentation System resulted in the cancellation of the next attempt scheduled for September 9.

On September 10, Jack Allavie and Charles Kuyk piloted the NB-52B to Silver Lake to launch White on his first Mach-3 flight. White's flight reached Mach-3.23 (2,182 miles per hour) and 79,864 feet altitude.

Commander Petersen made his second X-15 flight on October 20 1960. Fitzhugh Fulton and Charles Kuyk piloted the NB-52B. The flight plan for this flight was a repeat of the plan for Petersen's first flight. After turning west toward Mojave, he accelerated the X-15-1 to Mach-1.94 (1,280 miles per hour), climbing to a maximum altitude of 53,800 feet.

Eight days later, on October 28, NASA pilot Jack McKay made his first X-15 flight. Fitzhugh Fulton and Frank Cole flew the NB-52B. McKay made the usual pilot familiarization circuit from Palmdale to Boron, Mojave, and back to Edwards Air Force Base. He reached Mach-2.02 (1,333 miles per hour) and 50,700 feet. The Dryden Diary notation for this flight is "Pretty - Pretty".

Attempt to Launch Two X-15s in One Day

On November 3, 1960, the X-15-1 was mated to the NB-52B and the X-15-2 was mated to the NB-52A in an attempt to launch both of them on the same day. Bad weather set in and cancelled the attempt for that day.

On November 4, Scott Crossfield entered the cockpit of the X-15-2 to make an attempt at the first flight of the XLR-99 rocket engine. Captain Bob Rushworth got into the X-15-1 to make his first X-15 flight. Major Fulton and Major Cole piloted the NB-52A. After take-off, a hydraulic leak scrubbed Crossfield's flight. [36]

Major Fulton and Major Cole landed the NB-52A and climbed into the NB-52B to launch Captain Rushworth on his pilot familiarization flight. Rushworth reached Mach-1.95 (1,287 miles per hour) and an altitude of 48,900 feet.

Crossfield made the first flight of the X-15-2 with the XLR-99 engine on November 15. That flight and two more North American contractor demonstration flights of the XLR-99 equipped X-15-2 were launched from the NB-52A by early December. The X-15-2 was turned over to NASA for XLR-99 envelope expansion flights in early 1961. Joe Walker and Bob White would take turns pushing the X-15 to higher speeds and altitudes than it had reached before.

Neil Armstrong made his first X-15 flight on November 30. Frank Cole and Major Fulton piloted the NB-52B. The X-15-1 was launched over Palmdale. One of the rocket chambers of the XLR-11 engines shut down, so Armstrong completed the flight on seven cylinders. On his introductory flight, his top speed was just Mach-1.75 (1,155 miles per hour) and the highest altitude was only 48,840 feet.

The X-15-1 was fitted with the Q-ball nose in late 1960. Neil Armstrong was scheduled to make the first flight of the X-15-1 with the new nose on December 8, but the flight was postponed by weather until the following day. Major Allavie and Major Cole piloted the NB-52B. Neil Armstrong was launched over Palmdale. The primary objective of the flight was to evaluate the function of the new air data sensor, so Armstrong's top speed was only Mach-1.80 (1,188 miles per hour) and his maximum altitude was only 50,095 feet.

Replacement of the Movie Cameras

When the contractor phase of the X-15 flight test program was concluded in late 1960, North American retrieved the motion picture cameras that they had been using on the NB-52s. North American had been using a German Arriflex camera. The DBM #5 cameras used by the Air Force were somewhat larger and did not fit in the rear camera fairing on the NB-52s. One advantage of the DBM #5 was a built in heater, but the rear camera pod on the NB-52s had to be modified to accept the new camera in January of 1961. The modification was budgeted for $181.00 and 39.5 man-hours over three days.[37]

The mounts for the DBM #5 cameras were installed in the camera pods in February 1961. Their installation cast $244.00. Five days and 54.5 man-hours were required for the installation.

Inertial Flight Data System Pod

Inertial Flight Data System (IFDS) pod was developed that simulated the inertial navigation system of the X-15 to test the stable platform system of the NB-52B in flight without mating an X-15 to the Stratofortress. The pod was made from an external fuel tank left over from the North American F-107 Ultrasabre program. It fit on the pylon under the wing of the NB-52B. Pod flights were indicated by a special symbol painted on the fuselage of the NB-52B below the leading edge of the wing.

The pod was equipped with accelerometers to evaluate the forces that the X-15 was subjected to while the NB-52B taxied, took-off, and landed. Because the rear landing gear bogey of the NB-52B was so far behind the center of gravity, the nose of the Stratofortress dropped down hard. The X-15 was subjected to high g-loads when the nose gear of the NB-52 hit the runway. A slow-speed oscillograph was installed in the NB-52B for the pod accelerometer tests.

The X-15-1 was mated to the NB-52B in anticipation of Jack McKay's second X-15 flight, which was initially scheduled for January 11, 1961. The first launch attempt was aborted and the flight was rescheduled for January 26. Rains soaked the lakebed so the X-15-1 was de-mated.

The stable table pod replaced the X-15-1 on the pylon of the NB-52B for the first pod checkout flight on January 17. The pod was checked out on a second flight on January 27. Seven pod flights tested the complete inertial-platform system between March 1 and December 22, 1961.

Jack McKay's second X-15 flight was rescheduled for

The NB-52B taxis out with the X-15-1 before its sixteenth flight on November 4, 1960. Fitzhugh Fulton and Frank Cole piloted the Stratofortress. Bob Rushworth made his first flight in the X-15-1. (NASA Photo E6179)

The NB-52B with the X-15-1 sits in the foreground and the NB-52A with the X-15-2 sits in the background during an attempt to launch two X-15s in a single day on November 4, 1960. Due to a hydraulic leak, the attempt to launch Scott Crossfield on the first flight of the X-15-2 with the XLR-99 engine was aborted until November 15. The Day-Glo orange has been stripped from the engine nacelles of both Stratofortresses. (NASA Photo E-6186)

The NB-52B taxis out with the X-15-1 before its sixteenth flight on November 4, 1960. Fitzhugh Fulton and Frank Cole piloted the Stratofortress. Bob Rushworth made his first flight in the X-15-1. The tail of the NB-52A can be seen above the wing root. (NASA Photo E6189)

The inertial navigation systems of the X-15 were installed in this North American F-107A external fuel tank. The Inertial Flight Data System pod was carried on the X-15 pylon. (NASA Photo E-6261)

The inertial navigation systems of the X-15 were installed in this North American F-107A external fuel tan. (NASA Photo E-6262)

The NB-52B carrying the X-15-1 with the hot nose and XLR-11 engines sometime between December 1960 and February 1961. The Day-Glo orange has been temporarily stripped from the undersides of the wings, possibly at the same time that it was removed from the engine nacelles. Sometime before the next Open House on May 21, 1961, the left wingtip was painted black and a six-foot wide band of Day-Glo was painted near each wingtip. (AFFTC/HO)

January 31. Then it was postponed for another day by a hydraulic fluid heater failure. McKay was finally launched from the NB-52B over Palmdale on February 1. Major Fitzhugh Fulton and Major Ken Lewis flew the Stratofortress. McKay reached a top speed of Mach-1.88 (1,212 miles per hour) and a maximum altitude of 49,780 feet.

Last X-15 Flight with the XLR-11

The NB-52B launched the X-15-1 eleven times with the interim XLR-11 engines. The NB-52B launched the X-15-1 on the last flight with the XLR-11 engines on February 7, 1961. Major Fulton and Captain Robert Mosley piloted the Stratofortress. Stan Butchart manned the launch panel operator's station. Major Robert White was launched over Silver Lake, near Baker in eastern California. Through a combination of optimal conditions, the XLR-11 engines burned twelve seconds longer than they had on any prior flight. White flew the X-15-1 to Mach-3.49 (2,275 miles per hour), the highest speed that it attained with the XLR-11 engines.

First NASA XLR-99 Flight

The X-15-2 was scheduled to fly its first mission after its transfer to NASA on February 21, 1961. This flight marked the beginning of the envelope expansion phase of X-15 operations and the first flight of the X-15-2 with the hot nose. Major Fulton and

Major Robert White stands next to the X-15-1 after its twenty-first flight on February 7, 1961. It was the last X-15 flight flown with the interim XLR-11 rocket engine installation. (NASA Photo ECN-101)

Captain Kuyk piloted the NB-52B. Major Bob White was at the controls of the X-15-2. Before launch the operation of the stable table was interrupted when the APN-81 Doppler radar system shut down during the turn to the launch heading. The launch was aborted. Stable table troubles aborted a second launch attempt on February 24.

Following the mission abort on February 24, the X-15-2 was removed from the wing of the NB-52B and replaced by the IFDS pod. Captain Kuyk and Major Lewis piloted the NB-52B on the first test of the fully instrumented pod on March 1.

First Mach-4 Flight

The first NASA flight of the X-15-2 with the ball nose and XLR-99 engine was successfully launched over Silver Lake on March 7, 1961. Captain Kuyk and Major Frank Cole piloted the NB-52B. Jack Russell manned the launch panel operator's station. Major White piloted the X-15-2 to a new maximum speed of Mach-4.43 (2,905 miles per hour) and an altitude of 77,450 feet. It

Fitzhugh Fulton (standing right) and Robert Mosley (standing left) piloted the NB-52B on February 7, 1961. They launched Major Robert White in the X-15-1 on its twenty-first flight, represented by the last mission symbol in the bottom row. (Air Force Flight Test Center History Office)

By early 1961, the NB-52B had acquired nose art in the form of an eagle dropping an X-15 set against a blue circle. The eagle carried a ribbon in its beak bearing the Latin motto "Ut Viri Volent" (That man shall fly). The name "The Challenger" appeared under the eagle. The upper part of the blue circle was clipped. The badge of the Air Research and Development Command appears below the cockpit. Note that the lettering and badge are applied on bare metal rather than over Day-Glo orange paint. (Air Force Flight Test Center History Office)

was the first flight of any aircraft faster than Mach-4.

In March 1961, the ARA-25 radio in the NB-52B was returned to working order. If the X-15 pilot were to eject in the desert, the ARA-25 would allow the mothership to home in on the X-15 pilot's rescue beacon. The NB-52B could relay the position of the X-15 pilot to the high range stations on the telemetry frequency.[38]

Joe Walker was scheduled to make the fourteenth flight of the X-15-2 on March 21. The launch was cancelled due to "electrical trouble" in the NB-52B. When the NB-52B landed at the conclusion of the flight, the drogue chute separated from the tail of the Stratofortress. Major Fulton and Captain Allavie applied heavy braking to stop the NB-52B and its black payload on the runway. Several tires overheated. Their safety fuses blew out, deflating the tires.

On March 30, 1961, Joe Walker was launched from the NB-52B on the fourteenth flight of the X-15-2. Major Fulton and Captain Kuyk flew the Stratofortress. The XLR-99 engine failed to

Scott Crossfield, Bob White, and Neil Armstrong pose with the first two X-15s in 1961. Neil Armstrong flew an X-15 for the first time on November 30, 1960. Scott Crossfield made his last X-15 flight on December 6, 1960. (North American Rockwell)

Fitzhugh Fulton and Jack Allavie piloted the NB-52B on many missions. (Air Force Flight Test Center History Office)

Scott Crossfield, Bob White, and Neil Armstrong pose with the first two X-15s in 1961. The air-data booms of both X-15s have been replaced with ball noses. (AFFTC/HO)

The NB-52B carries the X-15-2 on February 21, 1961. (Greg Spahr)

The NB-52B carries Major Robert White in the X-15-2 on February 21, 1961. Major Fitzhugh Fulton and Captain Charles Kuyk piloted the Stratofortress. The X-15 flight was aborted due to problems with the "stable table". It was the first attempt to launch an X-15 equipped with the XLR-99 engine from the NB-52B. It was also the first flight of X-15-2 after installation of its "hot nose". (NASA photo E8698)

NB-52B carries Major Robert White in the X-15-2 on February 21, 1961. (NASA Photo E8698)

The NB-52B taxis with the X-15-2 on another aborted attempt to launch it on its thirteenth flight on February 24, 1961. Major Fitzhugh Fulton and Major Frank Cole piloted the Stratofortress. A horizontal x-15 mission mark has been added representing the aborted flight of February 21. The X-15-1 still wears the North American Aviation Markings. (AFFTC/HO)

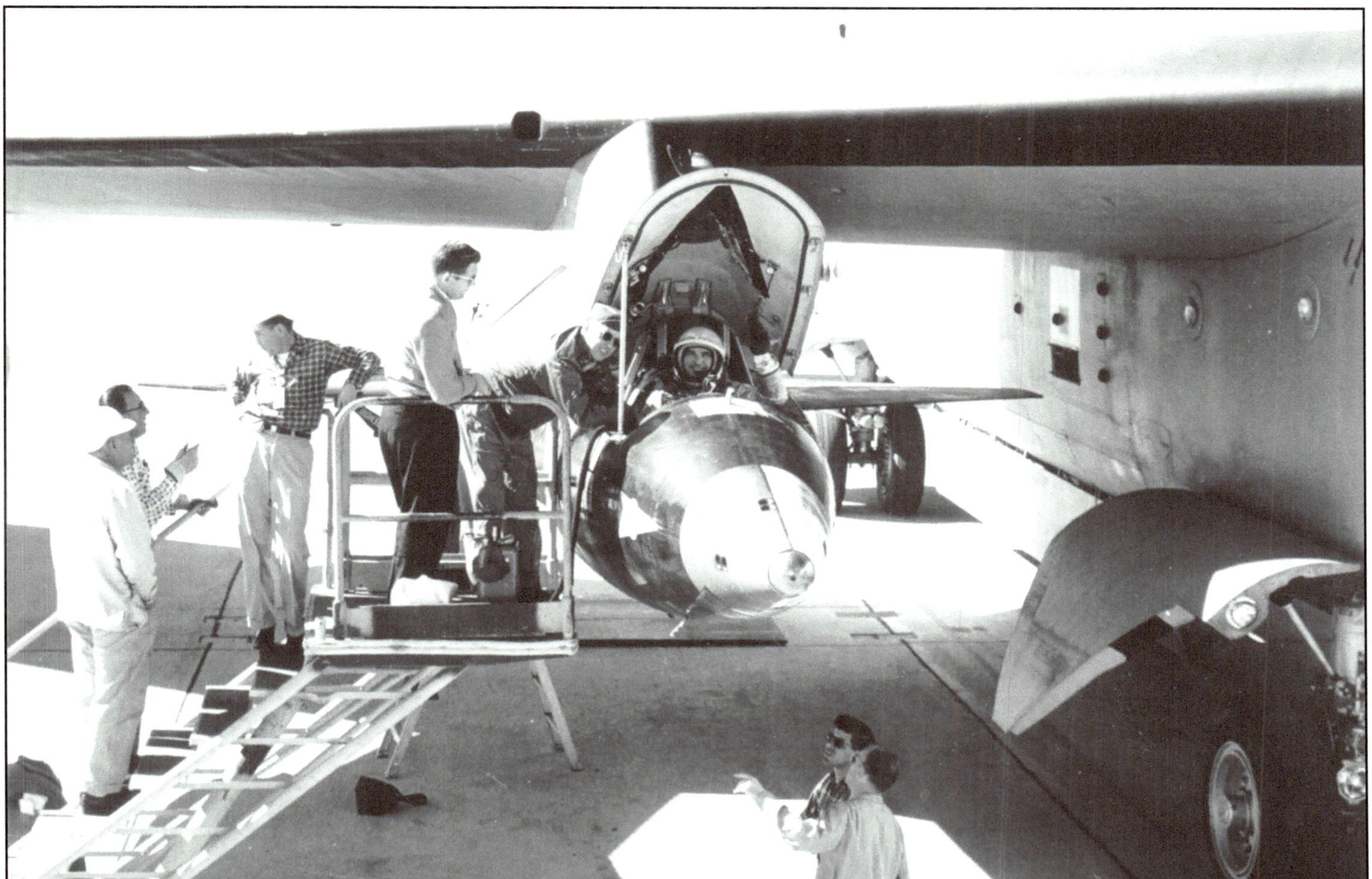

The X-15-2 hangs from the wing of the NB-52B later in 1961. The North American Aviation markings have been removed, but the NASA symbol on the nose is not complete. (AFFTC/HO)

Captain Jack Allavie and Captain Emil Sturmthal taxi the NB-52B past the audience at the 1961 open house. (American Aviation Historical Society negative library via Greg Spahr)

ignite on the first attempt. Walker reset the igniter, and it started on the second try. The X-15-2 reached an altitude of 169,000 feet.

Walker's flight reached an altitude over 30,000 feet higher than Robert White's record of 136,500 feet, set in the X-15-1 after launch from the NB-52A the previous August 12. This was to be the last absolute altitude record set by the X-15. Yuri Gagarin was launched on the first Vostok mission on April 12, eclipsing all previous absolute height and speed records.

Following a pod flight on April 4 it was discovered that the liquid nitrogen tank in the pylon had lost its insulating vacuum. The tank was removed and sent to North American for repairs.

Another pod flight was flown on April 5. The pod's accelerometers measured the g-forces that the X-15 would be subjected to as the NB-52B taxied to the runway. Captain Kuyk and Major Lewis flew the NB-52B on a simulated X-15 launch flight pattern. The launch panel operator observed how long it took to erect the stable table system. He evaluated three different modes of cooling the system. He checked out the operation of the transducer that provided the vertical velocity reference to the stable table. He determined how long it took the Doppler radar and other attitude sensing systems to recover from simulated launch pattern turns. He also evaluated the operation of the N-1 compass.

The next pod flight was scheduled for May 2. In April, the roll and pitch limit switches of APN-81 Doppler radar system were set to the maximum limit to alleviate the problem of the radar shutting down during the 180-degree turn to the launch heading.[39] Objectives for the flight included tests of modifications intended to reduce the recovery time of the stable table after the turn and evaluations of an improved temperature read-out and pylon cooling systems.

Shortly after brake release on the May 2 pod flight, a section of hot air duct failed on the number 7 engine in the right-hand, outboard nacelle. Escaping hot gasses activated the fire warning system, so the throttles were chopped, and heavy braking brought the NB-52B to a stop on the runway. Parts shortages delayed the repairs to the duct for several days.

The NB-52B was displayed in flight at the Edwards Air Force Base Open House on May 20 and May 22, 1961. Captain Jack Allavie and Captain Emil Sturmthal flew the Stratofortress past the audience.

Captain Allavie and Captain Sturmthal piloted the NB-52B on a pod flight on June 9. A recently installed 1,000-watt X-15 beacon checked out okay. The operation of the Doppler radar was found to be unsatisfactory. The tests were repeated on another pod flight piloted by Captain Kuyk and Captain Sturmthal on June 14. Captain Kuyk and Captain Sturmthal also piloted a pod flight on August 17.

MH-96 Adaptive Control System

The X-15-3 was tailored for flights to extreme altitude. For that purpose it was equipped with the Minneapolis-Honeywell MH-96 Adaptive Control System. The system adjusted the gain[5] of the aerodynamic flight control surfaces and blended the operation of those controls with the reaction control system on flights out of the atmosphere. The MH-96 had been developed for the Boeing X-20 Dyna-Soar.

As the X-15-3 ascended through air of decreasing density, the effectiveness of the control surfaces decreased as well. To get the same degree of control at higher altitude, the control surfaces needed to move farther for a given movement of the control stick. When the system sensed that the pilot was making large control movements, but the attitude of the X-15 was not responding adequately, the gain of the controls increased over time until the rocket plane reacted properly to the control inputs. When the gain reached 9/10s of the maximum value, the control surfaces

5 The amount of the movement of the control surfaces relative to the movement of the control stick is referred to as the gain.

The X-15-2 hangs from the wing of the NB-52B in 1961. (AFFTC/HO)

were nearly ineffective, and the system automatically engaged the reaction control thrusters.

As the X-15-3 reentered the atmosphere, the aerodynamic control surfaces became effective again. The MH-96 decreased the control gain to prevent the pilot from overcontrolling the descending spaceplane. When the gain fell to 65 percent, the system disengaged the reaction control thrusters, and the X-15 was controlled entirely by the aerodynamic controls for the rest of the flight.

The X-15-3 flew differently than the other two X-15s. As they other X-15s decelerated before landing, the pilot had to continually pull the stick back farther to keep the nose pitched up. The MH-96 system kept the attitude of the airplane steady when the pilot held the stick centered. As the airspeed and control surface effectiveness changed, the system adjusted the position of the control surfaces to maintain the pitch angle of the X-15-3. This was a different response than other aircraft of the time, but it simplified the task of maintaining the orientation of the X-15-3 during re-entry.

In June 1961 the NB-52s modified to incorporate the hardware associated with the operation of the MH-96 adaptive control system of the X-15-3. Control and switching units were added at the launch panel operator's station. An umbilical plug was added to the X-15 pylon to connect the MH-96 to the systems on board the Stratofortresses. An additional circuit breaker was installed for the MH-96 system. The cost of the installation was budgeted for $1010.00, and 160 man-hours were expended over 5 days.[40]

Ammonia Tank Outlet Temperature Readout

The XLR-99 engine used ammonia fuel in place of the water-alcohol fuel of the XLR-11 engines. In June 1961, a toggle switch was placed at the upper left corner of the Liquid Oxygen Top-off System console to allow the launch panel operator to switch the function of the engine lube oil temperature gauge on the console to read out the ammonia tank outlet temperature. Forty-eight man-hours and four days were required for the installation, which was budgeted for $313.00.[41]

The NB-52B flies over the X-15-2 sitting on Rogers Dry Lake following a mission in 1961. Two Air Force Lockheed F-104 Starfighter chase planes accompany the Stratofortress. (NASA Photo E14792)

APN-81 Doppler Radar Dummy Load Unit

It was discovered that the APN-81 Doppler radar unit was subjected to unacceptably high-level radar return signals during ground checks on the flightline. The signal reflected from the ramp overpowered the circuitry of the receiver unit of the Doppler radar. This was believed to have caused several component failures and system malfunctions.[42]

A dummy load unit was installed to absorb the energy of the Doppler radar transmitter while the NB-52B was on the ground.[43] The dummy load unit weighed three pounds and was mounted near station 450. The installation included a wave-guide and switch. Three days, thirty-nine man-hours, and $714.50 were budgeted for the installation.

The dummy load antenna switch was mounted on the APN-81 equipment rack. It was a two-position switch labeled "Dummy Load" and "Antenna". The Dummy Load position was to be used on the ground and the Antenna position was for airborne operation. The launch panel operator was cautioned that operation of the AN/APN-81 system on the ground might cause crystal damage reducing system performance sic drastically[44].

The installation of the dummy load unit had some unintended consequences. Subsequent to its installation, the dummy load switch was occasionally "inadvertently switched to dummy load by personnel moving about the aircraft" after the maintenance men had calibrated the radar and set the switch to the antenna position. The launch panel operator, once he was strapped into his seat in flight, couldn't get to the dummy load switch. On at least one occasion an X-15 flight was aborted because the switch was in the wrong position. An indicator light was added to the upper left corner of the launch panel operator's liquid oxygen top-off system console to show when the switch was in the dummy load position[45].

Joe Walker piloted the eighteenth flight of the X-15-2 on September 12, 1961. The recently promoted Major Jack Allavie and RAF Squadron Leader Harry Archer piloted the NB-52B. Jack Russell manned the launch panel operator's station.

Walker triggered the X-15 launch over Mud Lake, Nevada. The X-15-2 dropped away from the NB-52B and accelerated to 3,618 miles per hour at engine burnout. It was the fastest velocity achieved by the X-15 so far, but due to the flight profile and atmospheric conditions, the maximum Mach number attained was 5.21, slightly less than that achieved by Major White during the seventeenth flight of the X-15-2 on June 23. The maximum altitude was 114,300 feet.

Commander Forrest Petersen reached a new record Mach number of 5.3 on the next flight of the X-15-2 on September 28. Major Allavie and Royal Air Force Squadron Leader Archer flew the NB-52B to Hidden Hills Lake for the launch. Despite reaching a record Mach number, Petersen's highest speed was 3,600 miles per hour, slightly less than the speed that Joe Walker had achieved on the previous flight. Petersen reached a slightly lower maximum altitude of 101,800 feet.

Removing the Ventral Fin

The X-15 was designed with very large vertical stabilizers to insure stability at high speed. When flying at low angle of attack, the large surfaces had a favorable affect on roll-yaw coupling, but at high angle of attack, the large lower ventral stabilizer had an adverse affect on roll-yaw coupling. As the X-15 re-entered the atmosphere, the pilot pulled the nose up to a high angle of attack. During the pull-up, roll control inputs produced undesired movement around the yaw axis and vice versa.

As high-speed flight data was accumulated on X-15 flights, it was determined that removing the lower portion of the ventral fin improved controllability during re-entry. It was decided to fly the X-15 without the lower portion of the ventral fin.

Rushworth made the first flight of the X-15 without the ventral fin during the twenty-third flight of the X-15-1 on October 4, 1961. Starting with the twenty-ninth flight of the X-15-2 on September 28, 1962, almost all X-15 flights were made without the ventral fin. The exceptions were flights of the X-15A-2 with the ventral fin installed to provide additional surface area for tests of prospective heat protection materials.

First Mach-6 Flight

Captain Robert White exceeded Mach-6 for the first time on the twenty-first flight of the X-15-2 on November 9, 1961. Major Allavie and Squadron Leader Archer piloted the NB-52B.

After launch, Major White pulled up to a climb angle of twenty-five degrees and then pushed over for a zero-G acceleration run over the north end of Death Valley. The X-15-2 was going Mach 6.04 (4,093 miles per hour) when the engine burned out 83 seconds after launch. The maximum altitude of the flight was 101,600 feet.

For the second time in a month, one of the multiple panes of glass in the window of the cockpit shattered while Major White was flying the X-15-2. During the twentieth flight of the X-15-2, the outer, left panel of the left windshield cracked. A pane of soda-lime glass was inadvertently installed in place of the alumino-silicate glass that had been adopted for higher temperatures. This time it was the window on the right side of the cockpit canopy that failed.

Aerodynamic heating caused the Inconel-X window frame to buckle, pinching the glass panes. Conduction of heat from the frame generated thermal stresses that shattered the outer pane of the windshield. The alumino-silicate glass fractured into much

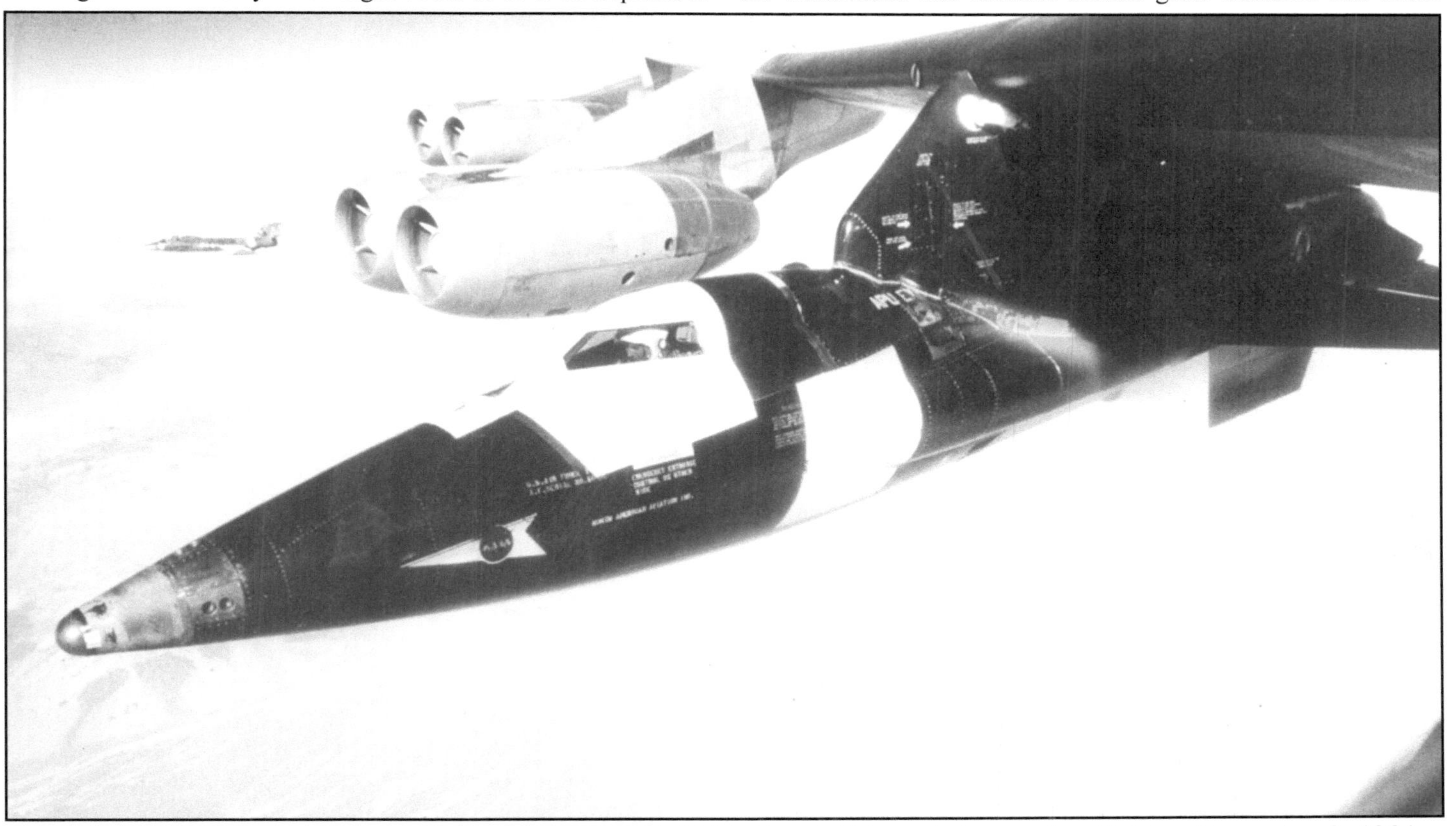

Captain Robert White shortly before launch from the NB-52B in the X-15-2 on its twenty-first flight on November 9, 1961. Major Jack Allavie and Squadron Leader Archer piloted the NB-52B. (North American Rockwell)

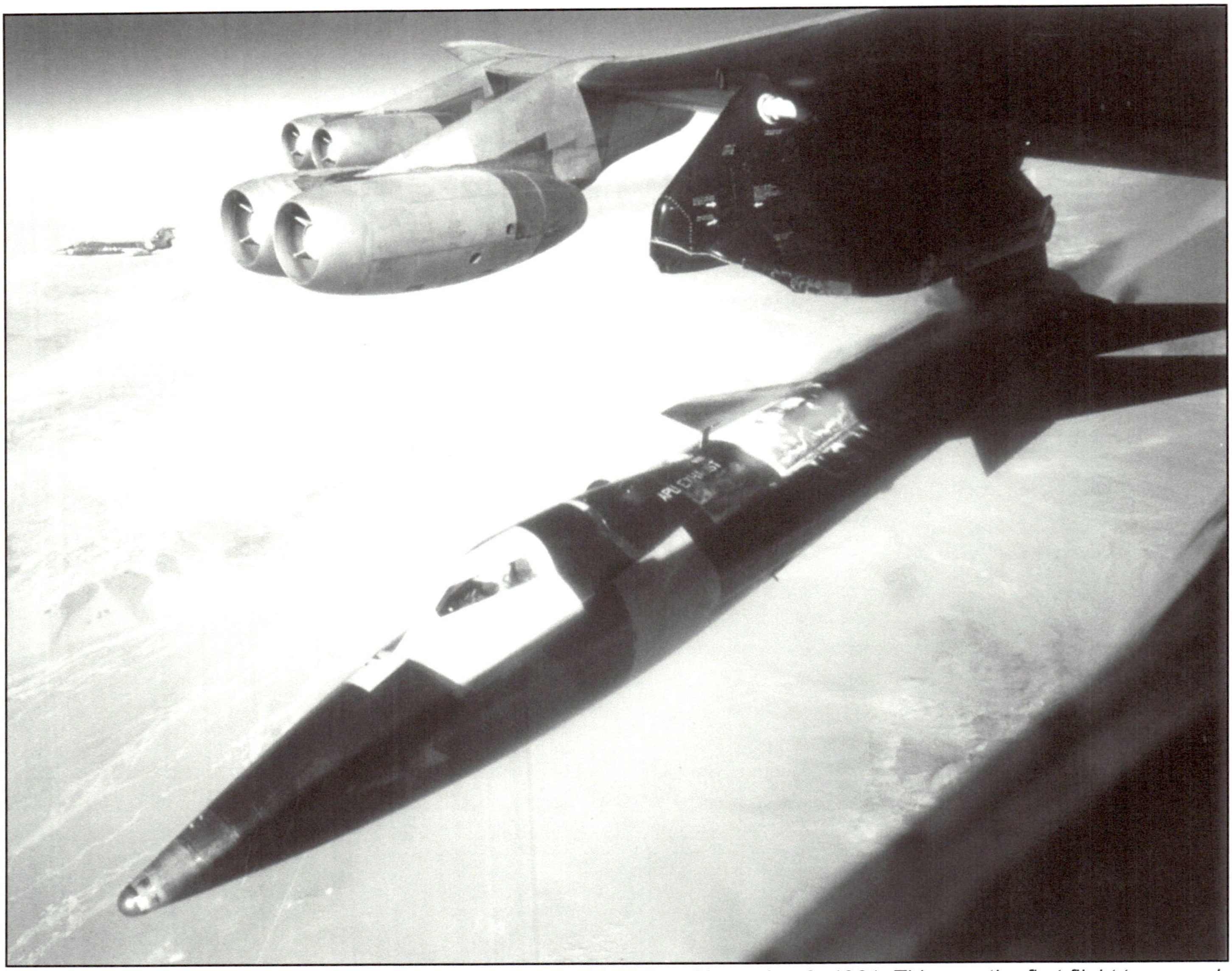

Captain Robert White is launched from the NB-52B in the X-15-2 on November 9, 1961. This was the first flight to exceed Mach 6, achieving Mach 6.04 (4,093 miles per hour) and an altitude of 101,600 feet. A Lockheed F-104 Starfighter is flying chase off the right wingtip. Patches of thermal protection paint have been applied to the left side of the cockpit canopy and the side of the fuselage. (North American Rockwell)

This picture was taken from the F-104 in the picture above on November 9, 1961. Note that the right side of the cockpit canopy was not painted white. The inner pane of the right windshield shattered due to thermal stresses on this flight. (AFFTC/HO)

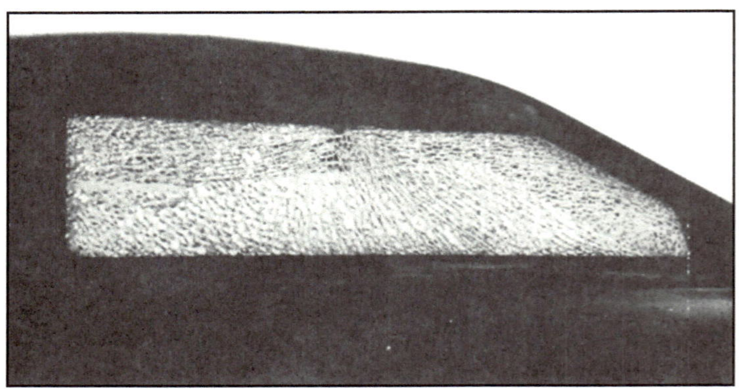

The inner panel of the right windshield of the X-15-2 shattered during its Mach-6 flight on November 9, 1961. The outer panel of the left windshield of the X-15-2 had shattered while Major White was piloting its twentieth flight on October 11. (NASA SP-60: X-15 Research Results)

X-15 mission symbols on the NB-52B shortly before the Stratofortress was repainted at Boeing in Wichita. The last mission symbol represents the twenty-first flight of the X-15-2, which was flown on November 9, 1961. Six "pod flight" symbols can be seen under the leading edge of the wing. (U. S. Air Force photo)

smaller pieces than the soda-lime glass window that had failed on the previous flight. The glass stayed in place, but the view on that side of the cockpit was obscured. If the window on the other side of the cockpit had broken, White would have been unable to see out of the cockpit. His only option would have been to fly on instruments until the approach to landing, then jettison the cockpit canopy and try to land the X-15 with an open cockpit.

The 0.05-inch thick Inconel X window frame was replaced with a 0.10-inch thick titanium frame to eliminate the buckling.

This flight concluded the speed envelope expansion for the X-15 in its original configuration. It had achieved the of Mach-6 set for the program nearly a decade earlier. The speed achieved on this flight would not be exceeded by an X-15 again for over two years.

In the course of the preceding five X-15 flights, the maximum Mach number had been increased three times and a new altitude record had been established. The only flight that did not set a record was flown to evaluate the stability of the X-15 with the lower ventral fin removed.

The movie *X-15* premiered in Hollywood over the Veteran's Day weekend. Staff from the Flight Research Center attended a screening on Friday November 10.

The last two IFDS pod flights were flown in December 1961. Captain Kuyk and frequent launch panel operator Stanley Butchart piloted a pod flight on December 13. Major Allavie and Major Cole piloted the last pod flight on December 22.

On January 2, 1962, Major Frank Cole and Captain Sturmthal flew the NB-52B to Boeing's facility at Wichita, Kansas for periodic maintenance. The exposed bare metal areas of the Stratofortress were painted silver. A new version of the nose art was placed lower on the nose so that the circle behind the eagle was not truncated at the edge of the Day-Glo orange. The last four digits of the Air Force serial number were painted on each side of the nose. The "U. S. AIR FORCE" lettering on the fuselage was changed from 36-inch tall letters to 24-inch tall letters and the Day-Glo on the nose was slightly truncated. At the conclusion of the maintenance, Captain Scharon and Major Kuyk flew the NB-52B back to Edwards Air Force Base on March 21.

The Longest X-15 Flight

Neil Armstrong inadvertently flew the longest X-15 mission during the fourth flight of the X-15-3 on April 20, 1962. Major

The NB-52B is prepared for an X-15 launch in April 1962, shortly after returning from repainting at Boeing. The Day-Glo orange around the cockpit is free from weathering. (North American Rockwell)

The NB-52B taxis with the X-15-2 on an aborted attempt to launch it on its twenty-third flight on May 25, 1962. Major Fulton and Major Bement piloted the Stratofortress. Major White occupied the cockpit of the X-15-2. The flight was aborted due to overheating of the "stable table". The Day-Glo orange around the cockpit is already badly chipped, less than two months after the Stratofortress returned from repainting at Boeing. (AFFTC/HO)

The NB-52B taxis with the X-15-2 before its twenty-fourth flight on June 29, 1962. Major Allavie and Squadron Leader Archer piloted the Stratofortress. Jack McKay occupied the cockpit of the X-15-2. (AFFTC/HO)

Jack Allavie and Major Russell Bement were at the controls of the NB-52B. Stanley Butchart manned the launch panel operator's station. There was severe turbulence at the launch altitude. The pitching and rolling of the NB-52B interfered with attempts to get a solid fix from the Doppler radar. The launch proceeded under the assumption that the velocity data supplied to the MH-96 aboard the X-15-3 would be slightly in error.

Armstrong got a good light of the XLR-99 engine on the first try. The flight followed the flight plan closely during the ascent. Armstrong made a series of pull-ups, evaluating the stability augmentation system at angles of attack up to 20 degrees. The X-15-3 was going mach 5.31 when he shut down the rocket engine eighty-one seconds into the flight. It reached a peak altitude of 207,500 feet.

Armstrong felt that the aircraft control and damping during re-entry were better than expected. The MH-96 system was using reaction control thrusters more than had been expected. That was depleting the hydrogen peroxide that fueled both the thrusters and the auxiliary power units. The peroxide-low warning lamp came on as the X-15 descended through 160,000 feet. Armstrong switched on the pump that transferred unused hydrogen peroxide from the engine turbopump to the tank that supplied the thrusters and auxiliary power units.

Although the flight plan had called for Armstrong to level out at 90,000 feet, he inadvertently pulled the nose up too high. The X-15-3 skipped back up to an altitude where its wings produced almost no lift or drag. Rather than decelerating, the X-15 continued at nearly unabated speed toward Edwards Air Force

The NB-52B taxis with the X-15-2 before its twenty-fourth flight on June 29, 1962. The last X-15 mission symbol represents the sixth flight of the X-15-3, which was flown on June 21. (AFFTC/HO)

The X-15-2 is prepared for launch from the NB-52B on July 19, 1962. Jack McKay piloted the twenty-fifth flight of the X-15-2. (AFFTC/HO)

The X-15-2 is prepared for launch from the NB-52B on July 19, 1962. Major Fulton and Major Bement piloted the Stratofortress. The last X-15 mission symbol represents the thirty-first flight of the X-15-1, which was flown on July 16. (AFFTC/HO)

Jack McKay prepares the X-15-2 for launch on its twenty-fifth flight on July 19, 1962. (AFFTC/HO)

Base. Jack Allavie in the NB-52B could see that Armstrong was passing Rogers Dry Lake and called out, "Hard left turn, Neil", over the radio. Armstrong rolled the X-15 onto its left side and pulled back on the stick until the stabilizers were deflected as far as they could go, but the X-15 continued in a nearly straight line. He watched his landing site pass behind him as he was still traveling Mach 3 and heading for Los Angeles.

When Armstrong got the X-15-3 turned around he was many miles south of the base. He was flying over the southern San Gabriel Mountains, far from his planned flight course. The sonic boom from the X-15 swept across La Canada and the northern Los Angeles basin.

Armstrong had the impression that he was "in pretty bad shape for the south lakebed". For a moment he considered putting the X-15 down on the concrete runway at the Palmdale Airport. He pitched the X-15 to the angle of attack that generated the most lift for the least drag to stretch his glide as far as he could. He didn't bother with the usual spiraling descent. He made a straight in approach to Runway 35 on the south lakebed.

The X-15 pilots had not considered the prospect of landing on Rogers Dry Lake from the south. There were no pre-selected checkpoints to help guide Armstrong to his landing point. Late in the approach, he realized that he would be short of his initial aim point by about two miles. He was flying below the tops of the Joshua Trees before he reached the lakebed. The X-15 landed on the hard clay surface just a short distance from the sagebrush. The X-15 touched down 12 minutes and 28 seconds after launch, nearly a minute and a half longer than the original flight plan. This was the longest flight of the X-15 program.

Armstrong had missed his planned landing site by twelve miles. The convoy of recovery vehicles was at the far end of Rogers Dry Lake and did not arrive for a quarter hour.

Altitude record – 255,000 feet

The next X-15 flight was an altitude build up flight piloted by Joe Walker in the X-15-1. Walker's flight plan called for him to establish a climb angle of 38 degrees after launching over Mud Lake. He was to let the rocket engine burn at full throttle for 81 seconds. Paradoxically, Walker was instructed to extend the speed brakes during the engine burn above 70,000 feet. This improved the directional stability of the X-15 in the rarefied upper atmosphere. The X-15 was expected to be going about Mach 5.35 at engine burn out, after which it would coast to a peak altitude of 255,000 feet. During the descent, Walker would begin his pullout at 180,000 feet and level out at 70,000 feet, 20 miles lower. Walker planned to land at Rogers Dry Lake about 11 minutes after launch.

Walker's flight had already been postponed once when Major Allavie and Squadron Leader Archer took the NB-52B aloft with the X-15-1 on April 27. Cloudy skies over the launch lake forced another scrub of the launch that morning. The launch was rescheduled to April 30. Major Russell Bement was in the co-pilot's seat of the NB-52B for the next launch attempt. Jack Russell monitored the systems of the X-15 at the launch panel operator's station.

The XLR-99 engine lit up without delay and Walker established his climb. The flight progressed just as planned. Walker found that he had to strain to reach the throttle to shut off the engine under the force of nearly 3 Gs of acceleration. The X-15 was traveling Mach-4.73 (3,489 miles per hour) at engine shut down and continued to climb another 20 miles, reaching a peak altitude of 247,000 feet about 80 seconds later. As Walker pushed over at the peak altitude, he got a strong feeling that he was going to overshoot Edwards Air Force Base. The maximum Mach number, Mach-4.94, was achieved during the descent.

The staff of the NASA Flight Research Center headed over to the Elks club for a party that night to celebrate the new altitude record that Walker had set.

Major Rushworth was scheduled to fly the X-15-2 on May 7, but overcast skies postponed the launch for a day. The NB-52B took off on May 8 with Major Allavie and Major Russell Bement at the controls. Jack Russell occupied the launch panel operator's station behind the pilots. The number four J57 jet engine malfunctioned and was shut down. The flight progressed on the power of the Stratofortress' other seven engines.

Rushworth triggered the drop of the X-15-2 over Hidden Hills Lake and started climbing at full throttle. He leveled off at 70,000 feet and throttled the XLR-99 engine to 30% thrust. This was the first time that the XLR-99 had flown at less than 50% thrust. Rushworth extended the speed brakes 16 degrees to maintain a speed of Mach-5 for 33 seconds. The edges of the speed brake panels were heated to a temperature of 1,250 degrees Fahrenheit by aerodynamic friction. The pressure in the cockpit of the X-15 dropped, and Rushworth's pressure suit inflated just as the rocket engine consumed the last of its fuel. The X-15 was traveling Mach-5.34 (3,524 miles per hour) at engine burn out. Rushworth rolled the X-15 ninety degrees to the right and pulled back on the stick until he was subjected to 4.5 gs. Rushworth was impressed by the "snap, crackle, and pop" sounds emanating from the X-15 as it glided back to Edwards Air Force Base.

The 1962 Edwards Air Force Base Open House was held on May 20. Major Allavie and RAF Squadron Leader Harry Archer piloted the NB-52B as it participated in the rehearsal for the aerial display on May 19. The NB-52A with the X-15-3 mounted on its wing was placed on static display on the flightline during the airshow. Major Miles Burganheim occupied the co-pilot's seat of the NB-52B as Major Allavie flew it past the crowd on May 20.

Major Bob White was scheduled to make the twenty-third flight of the X-15-2 on May 24. The primary purpose of the flight was to evaluate the performance of the alternate stability augmentation system. The flight plan called for White to pull up to a climb angle of 30 degrees and climb at full throttle. At 56,000 feet White would push over until the wings of the X-15 produced no lift. He was

The NB-52B takes off with an X15 in the Summer of 1962. (United States Air Force Museum)

NASA Flight Research Center pilots pose in front of the Bell X-1E on October 2, 1962. From left to right Neil Armstrong, Joe Walker, Bill Dana, Bruce Peterson, Jack McKay, Milt Thompson, and Stan Butchart. Armstrong, Walker, Dana, McKay and Thompson flew the X-15s. Peterson later flew the HL-10 and Mr-F2 lifting bodies Butchart had piloted the Boeing Superfortresses that launched the Bell X-1s and Douglas D-558-II Skyrockets and often served as launch panel operator on early X-15 missions. (NASA Photo EC-149)

The X-15-2 drops away from the NB-52B sometime between June 1 and November 9, 1962. (NASA Photo EC88-108-1)

to continue flying a zero-g ballistic trajectory for forty seconds. Seventy-five seconds into the flight he was to pull up so that the X-15 was pulling 2 Gs for the last eighteen seconds of powered flight. After shutting down the XLR-99 engine, White was to pull up to an angle of attack of 23 degrees, disengage the roll channel of the primary stability augmentation system and observe the behavior of the X-15 as the alternate stability augmentation system kicked in. Then he would re-engage the stability augmentation system and roll the X-15 thirty degrees to the right while holding the angle of attack at 20 degrees for twenty seconds during the ascent. The X-15 should reach a peak altitude of 162,000 feet three minutes after launch.

Bad weather on the morning of May 24 resulted in a postponement of the flight until the next day. Major Fitzhugh Fulton and Major Russell Bement took the NB-52B aloft with White in the cockpit of the X-15-2 on May 25. As Jack Russell was setting up the inertial navigation system, the Stable Table Overheat light came on, causing the mission to be rescheduled for May 28. It was postponed again to May 29, but the stable table malfunctioned again that day. This was the fourth consecutive launch abort for Major White.

Major Fitzhugh Fulton and Major Russell Bement successfully launched the twenty-third flight of the X-15-2 over Delamar Lake on June 1, 1962. Major White piloted the X-15-2 to a speed of Mach-5.43 (3,675 miles per hour) and an altitude of 132,600 feet.

Major Russell Bement and Alvin White flew the NB-52B past the Palmdale Airshow on October 28, 1962.

X-15-2 Landing Accident

The X-15-2 was involved in a serious accident on its thirty-first flight on November 9, 1962. Major Bement and Lt. Col. Kenneth Lewis launched Jack McKay in the X-15-2 from the NB-52B over Mud Dry Lake. It was the fourteenth flight of the X-15-2 that was launched from the NB-52B since the installation of its XLR-99 engine. The XLR-99 engine throttle stuck at 30% thrust.

Jack McKay had to make an emergency landing on the lakebed. He was unable to jettison all of the propellants before landing and the flaps failed to deploy. His landing speed was 296 miles per hour instead of the usual 230. Just after touchdown, the left main landing gear failed. The left wing tip dug into the surface of the dry lake and flipped the X-15 onto its back. McKay jettisoned the cockpit canopy to avoid being trapped in the cockpit. The top of his helmet hit the surface of the lake before the fuselage of the X-15.

McKay suffered a compressed spine and internal injuries.

The severed left skid of the X-15-2 lies on Mud Dry Lake after Jack McKay's landing accident. (AFFTC/HO

The X-15-2 lies on its back on Mud Dry Lake following Jack McKay's landing accident on November 9, 1962. (AFFTC/HO)

A little while later the ice has melted on the X-15-2 as it lies on its back on Mud Dry Lake following Jack McKay's landing accident on November 9, 1962. (AFFTC/HO)

The X-15-2 lies on its back on Mud Dry Lake following Jack McKay's landing accident on November 9, 1962. (AFFTC/HO)

He was unable to extricate himself from the inverted rocket plane. His lungs were damaged by the vapors of anhydrous ammonia escaping from the fuel tank.

Astronaut Wings

The NB-52B launched X-15 flights that qualified five pilots as astronauts by exceeding an altitude of fifty miles.

Joe Walker became the first civilian astronaut when he flew the X-15-3 to an altitude of 271,700 feet during its fourteenth flight on January 17, 1963. Major Bement and Squadron Leader Archer launched Walker from the NB-52B over Delamar Lake. Stanley Butchart manned the launch panel operator's station. The X-15-3 was going Mach-5.18 at engine shutdown, but it reached Mach-5.47 as it descended from the peak altitude.

Major Robert Rushworth was the second pilot to earn his astronaut wings after launch from the NB-52B on June 27, 1963. Major Bement and Squadron Leader Archer launched Rushworth in the X-15-3 on its twentieth flight over Delamar Dry Lake. Stanley Butchart manned the launch panel operator's station. Rushworth reached 280,000 feet. The X-15-3 was going Mach-4.76 at engine shutdown, but it reached Mach-4.89 as it descended from the peak altitude.

North American X-15A-2

While the X-15-2 was being repaired, the opportunity was taken to modify it for even higher speeds. By raising the top speed of the X-15 to over Mach-7, it would become possible to test supersonic combustion ramjets. To accomplish this mission, it would need additional fuel for a longer engine burn time, and tanks and fixtures to support the operation of the ramjet. To reflect the modified status of the high-speed variant of the X-15, it was re-designated as the X-15A-2.

The most apparent modification was the addition of large external fuel and oxidizer tanks. The fuel in the external tanks added seventy percent to the available engine burn time. The right hand external tank carried 1,080 gallons of anhydrous ammonia. The left-hand tank could carry 793 gallons of liquid oxygen. The left tank also contained three high-pressure helium bottles. The helium was used to pressurize the propellants and force them into the internal tanks of the X-15. The left tank weighed over two thousand pounds more than the right tank when they were fully loaded. Together they added 15,770 pounds to the weight of the fully fueled X-15A-2. The external tanks contained 13,500 pounds of propellants.

Both tanks were equipped with gas cartridges to eject them from the X-15A-2. The gas cartridges were mounted on the fore and aft ends of the tanks. A solid rocket motor was mounted on the front end of both tanks to force them away from the X-15 at the moment they were jettisoned. All of the cartridges and the solid rocket motors would be fired to jettison the tanks when their fuel had been exhausted. If the pilot of the X-15 failed to get the engine to light, the full tanks would be jettisoned by firing only the front gas cartridges.

The X-15-2 at North American being rebuilt following Jack McKay's accident in November 1962. Note the missing right wingtip. It was replaced with a removable wingtip for the possible installation of alternative airfoils.(North (American Rockwell)

Flutter evaluation tests were conducted with a scale model of the NB-52 to simulate the effect of the heavier X-15A-2. (Boeing)

The external fuel tanks were reusable. Each contained a parachute recovery system with two parachutes. The first parachutes, small drogue chutes, were deployed immediately after the tanks were jettisoned to keep them from tumbling and stabilize their descent. The main parachutes were triggered at an altitude of 8,000 feet by an altimeter. The need for a suitable range for the impact of the tanks placed further constraints on the locations considered for the launch of the hypersonic vehicle.

To make room for the ramjet's liquid hydrogen fuel tanks, the fuselage of the X-15A-2 was stretched by twenty-nine inches. The additional fuselage section was inserted between the liquid oxygen and ammonia tanks near the center of gravity of the airframe.

The extra engine run time required more hydrogen peroxide

Fitzhugh Fulton poses in front of the NB-52B as the X-15-3 is prepared for flight. The last visible X-15 mission symbol represents an aborted attempt to launch the X-15-1 on its thirty-seventh flight on July 3, 1963. (AFFTC/HO)

to drive the turbopump of the XLR-99 engine. Additional hydrogen peroxide tanks were installed in extensions of the fuselage side fairings. An additional helium bottle was installed above the engine, aft of the vertical stabilizer, to augment the gas supply for pressurizing the propellants.

A new lower fin was designed to mount a supersonic combustion ramjet test shape. The full-size ventral stabilizer could be attached for tests of ablative heat-protection material.

The main landing gear legs of the X-15A-2 were lengthened to allow it to land with the test ramjet mounted on the ventral fin. Intermediate length main gear legs, 6.75 inches longer than the original legs, were fabricated for flights without the ramjet. The shorter legs imposed less strain on the fuselage when the nose gear impacted the ground.

The windshield was redesigned in anticipation of the greater heat induced stresses that would be imposed by the higher speeds that the X-15A-2 would achieve. To avoid the concentration of stresses at the corners of the original trapezoidal windows, they were replaced with oval windows. The two-pane construction of the original windshield was revised to incorporate three panes of alumino-silicate glass.

It was decided to adopt the use of a coating for the X-15 that would protect it at the higher temperatures of Mach-7 flight. The material would protect the metal structure of the X-15 by ablating. At high temperatures, the ablative material would undergo a chemical reaction that absorbed heat, reducing the rate of heating of the underlying Inconel-X panels. Candidate materials were to be tested as panels mounted on the X-15A-2 in a variety of locations.

A "sky hatch" was added to the experiment bay behind the cockpit. The sky hatch could be opened by the pilot of the X-15 to expose experiments in the bay to the sky. This feature was used for an ultraviolet stellar photography experiment.

The damaged right wingtip was replaced with a detachable wingtip. Experiments were envisioned in which the wingtip would be replaced with alternative airfoils or structures. [46]

The stretch of the fuselage of the X-15A-2 and the addition of the external tanks increased its launch weight from 31,000

NB-52B carries the X-15-3 before launch on its twenty-first flight on July 19, 1963. Lt. Col. Fitzhugh Fulton and Major Russell Bement piloted the Stratofortress. Joe Walker flew the X-15-3 to a new altitude record of 347,800 feet. Captain Albert Crews piloted the Air Force Northrop T-38A Talon chase plane. (AFFTC/HO)

Milton Thompson stands next to the X-15-3 following a flight in 1964. (NASA photo ECN-413)

pounds to over 50,000 pounds. The heaviest payload ever carried by the NB-52B was the X-15A-2, weighing 53,100 pounds.

Major Robert Rushworth was scheduled to make the twenty-third flight of the X-15-3 on October 14, 1963. Lt. Col Fulton and L. Col. Kenneth Lewis piloted the NB-52B. Jack Russell manned the launch panel operator's station. Problems with the stable table resulted in an aborted mission. An attempt to fly the mission on October 15 was cancelled due to a stuck LOX valve. Bad weather postponed attempts on the following two days.

The twenty-third flight of the X-15-3 was rescheduled for October 24. As the NB-52B taxied out to the runway, one of the panes of its windshield shattered, and the mission was cancelled again. The windshield was replaced in time to make another launch attempt on October 25. Lt. Col Fulton and Major Charles Bock flew the Stratofortress. Stanley Butchart manned the launch panel operator's station. The launch preparations were held up for one

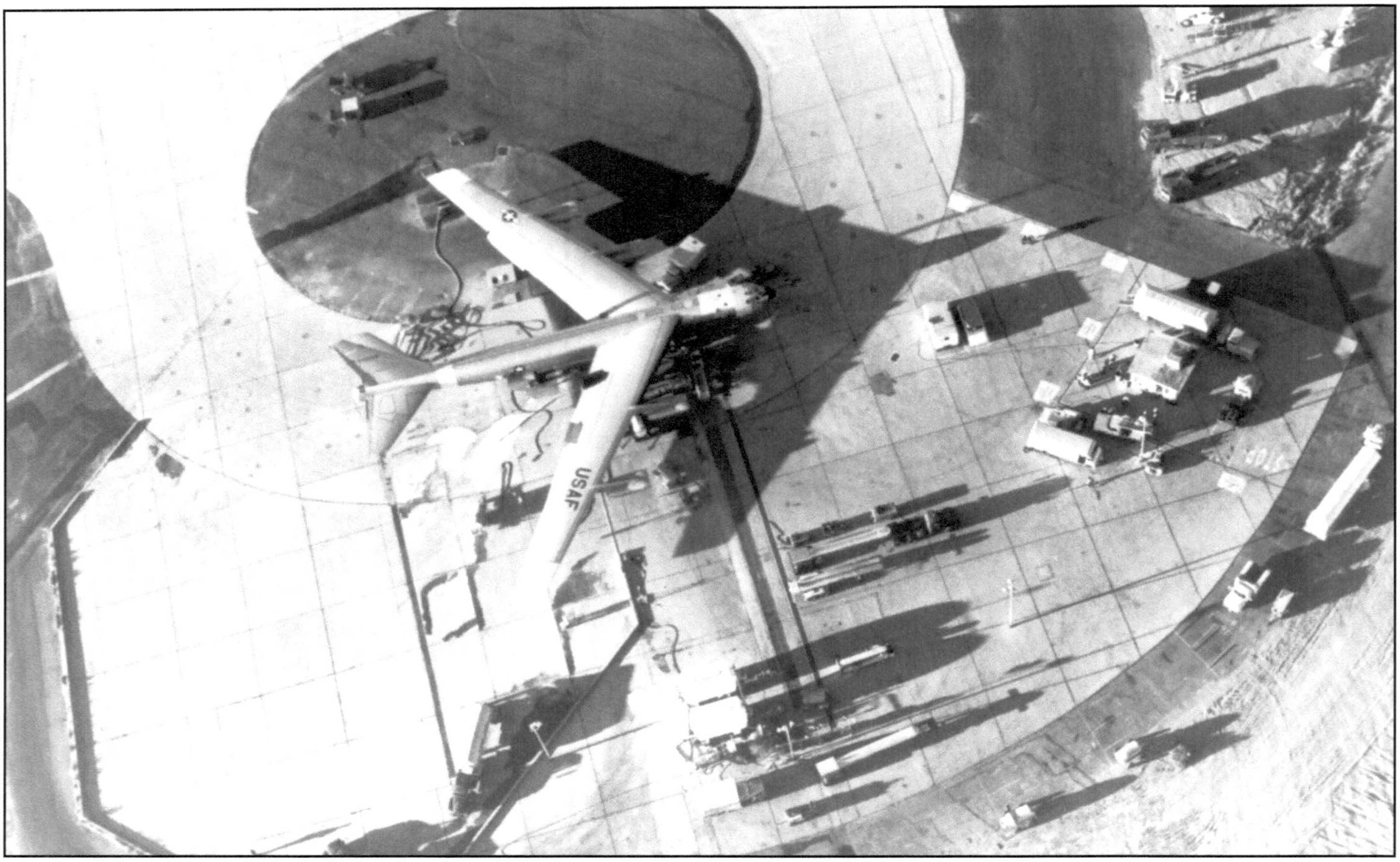

The NB-52B is prepared for an X-15 mission in 1963 or 1964. Sprays of water on the concrete can be seen behind the wing of the NB-52B. Overflows of Hydrogen peroxide and Liquid Oxygen are directed into the water sprays to prevent any ignition. (AFFTC/HO)

hour and forty-five minutes by a source pressure leak. Once aloft, the flight was aborted by problems with the stable table.

The X-15-3 was removed from the wing of the NB-52B and the X-15-1 took its place. Lt. Col Fulton and Major Gayland Jones successfully launched Milton Thompson on his first X-15 flight on the first attempt on October 29. Stanley Butchart manned the launch panel operator's station. Thompson piloted the X-15-1 to a speed of Mach-4.10 (2,712 miles per hour) and an altitude of 74,400 feet.

Major Jones and Major Bement finally launched Rushworth on the twenty-third flight of the X-15-3 over Hidden Hills Lake on November 7. Rushworth piloted the X-15-1 to a speed of Mach-4.40 (2,925 miles per hour) and an altitude of 74,400 feet.

Major Bement and Major Gayland Jones flew the NB-52B past the Palmdale Airshow on November 9, 1963.

On January 8, 1964, Captain Joe Engle was scheduled to be launched from the NB-52B by Lt. Col. Fulton and Lt. Col. Lewis on the forty-third flight of the X-15-1. Fifteen seconds before launch, the outboard engine of the nacelle next to the X-15 started spitting out molten metal. The chase pilot advised Lt. Col. Fulton of the engine failure. Fulton stated that he was not going to shut the engine down until the X-15 had been launched.[47] Once the X-15 had been dropped, Fulton shut down both engines in that nacelle. The NB-52B returned to Edwards Air Force Base on the power of the remaining six engines.[48]

After the completion of the rework of the X-15A-2 by North American in El Segundo, the spaceplane was trucked back out Sierra Highway to Edwards Air Force Base on February 18, 1964.[49]

The pylon of the NB-52B was modified to carry the heavier X-15A-2. The modifications were completed in April 1964.

On April 17, 1964, Major Bement and RAF Flight Leader David Cretney flew the NB-52B to Tinker Air Force Base in Oklahoma by for Inspection and Repair as Necessary (IRAN). [50] While the NB-52B was at Wichita, the last of the Day-Glo orange was removed from the airframe and the engine nacelles were painted light gray.

IRAN for the NB-52B was completed a month ahead of schedule. Major Bement and Major Saunders flew a local maintenance check flight at Tinker Air Force Base on July 22, 1964. Major Bement and Major Bock flew the NB-52B back to Edwards Air Force Base the following day.

On August 14, following launch from the NB-52A, Major Rushworth felt the X-15A-2 shudder and begin to vibrate as it decelerated through Mach-4.4. The nose landing gear had deployed. Aerodynamic heating cause the fuselage to grow longer. The landing gear release lanyard incorporated a bit of slack to accommodate the increase in length of the fuselage. Due to the

The NB-52B received new nose art depicting a B-52 throwing an X 15 after returning from IRAN at Tinker Air Force Base in the summer of 1964. (NASA photo E12089)

extra length of the X-15A-2, the slack in the landing gear release lanyard was no longer sufficient to prevent the inadvertent release of the landing gear. Despite the destruction of the nose gear tires, Rushworth landed the X-15A-2 safely on Rogers Dry Lake.

The NB-52B first launched the modified X-15A-2 on September 29, 1964. Lt. Col. Fulton and Col. Guy Townsend piloted the Stratofortress. Stanley Butchart manned the launch panel operator's station. Major Rushworth piloted the third fight of the X-15A-2 in its new configuration. The X-15A-2 reached Mach-5.20 (3,542 miles per hour) and an altitude of 97,800 feet. Despite a redesign of key linkages in the nose gear, the small door in the center of the nose gear door popped open during the flight.

The NB-52B made appearances at both days of the 1964 Pt. Mugu Naval Weapons Center airshow. Col. Joseph Cotton piloted the NB-52B on both days. Major Jones flew as co-pilot on October 17, and Major Bock accompanied Col. Cotton on October 18.

Two weeks later, on November 1, Major Bement and Major Jones flew the NB-52B past the Palmdale Airshow.

Jack McKay Flies the X-15 Again

Jack McKay returned to the cockpit of the X-15A-2 during two captive-carry flights on the NB-52B in November 1964. Lt. Col. Fulton and Major Bock piloted a captive-carry mission flown on November 6. Major Bement and Col. Cotton piloted a second captive-carry mission flown on November 16.

Major Bement and Major Bock launched Jack McKay from the NB-52B in the X-15A-2 on November 30, 1964. McKay's first flight in the modified rocket was limited to Mach-4.66 to check

The X-15-1 hangs from the pylon of the NB-52B before its fifty-second flight on February 26, 1965. Lt. Col. Fitzhugh Fulton and Major Charles Bock piloted the Stratofortress. Jack Russell manned the launch panel. Jack McKay piloted the X-15-1 to a speed of Mach-5.4 (3,702 miles per hour). (NASA Photo E13049)

X-15 pilot Jack McKay poses with the X-15A-2 on December 2, 1965. (North American Rockwell)

NB-52B takes off with X-15-1 on February 26, 1965. Fitzhugh Fulton and Charles Bock launched Jack McKay on a flight that reached a speed of Mach-5.4 and an altitude of 153.600 feet. (NASA photo E13057)

The NB-52B carries an X-15 in 1965. (Air Force Flight Test Center History Office)

The NB-52B takes off from Runway 04 with the X-15A-2 in 1965. (NASA Photo ECN722)

The NB-52B carries the X-15-3 before its fiftieth flight on October 12, 1965. Lt. Col. Fulton and Major Gayland Jones piloted the Stratofortress. Jack Russell manned the launch panel. Captain William "Pete" knight piloted the X-15-3. (AFFTC/HO)

out the modifications that had been made to the nose gear door to prevent unintended gear release in flight. [51]

The NB-52B carried the X-15A-2 on a captive flight on February 15, 1965. Major Bement and Lt. Colonel Lewis piloted the NB-52B and Major Rushworth occupied the cockpit of the X-15A-2. The photographer in the chase plane failed to get photographs of the X-15A-2, so the mission was repeated that afternoon. Major Bock took over as co-pilot of the NB-52B for the second mission.

X-15A-2 Main Gear Deploys at Mach-4.4

On February 17, 1965, Lt. Colonel Fulton and Major Bement launched Major Rushworth in the X-15A-2 near Mud Lake. Rushworth piloted the X-15A-2 to a top speed of Mach-5.27 (3,539 miles per hour). Once again, as he decelerated through Mach-4.4, he heard a bang and felt the X-15A-2 shudder. On his third consecutive flight in the X-15A-2, the landing gear of the X-15A-2 extended at high speed. The chase pilot that intercepted the X-15A-2 at Edwards Air Force Base reported to Rushworth that the right main landing gear had deployed, but that it appeared to be structurally sound. Rushworth's landing on Rogers Dry Lake was unremarkable considering the fact that his landing gear had deployed while the X-15A-2 was moving over two thousand, five hundred miles per hour. [52]

Col. Cotton and Major Bement flew the NB-52B past the audience at the Edwards Air Force Base open house on May 16, 1965.

Captain Joe Engle became the third pilot to qualify for astronaut wings after launch from the NB-52B on June 29, 1965. Lt. Colonel Fulton and Lt. Colonel Andonian launched Engle in the X-15-3 on its forty-fourth flight over Delamar Dry Lake. Jack Russell manned the launch panel operator's station. Engle reached an altitude of 280,600 feet. The X-15-3 was going Mach-4.84 when the XLR-99 engine shut down, but it reached Mach-4.94 on the descent from the peak altitude.

The NB-52B carries the X-15-3 before its fiftieth flight on October 12, 1965. (Air Force Flight Test Center History Office)

The NB-52B carries the X-15-3 on October 12, 1965. Fulton and Jones launched Knight on a flight that reached a speed of Mach-4.62 (3,118 miles per hour) and an altitude of 94,400 feet. (Air Force Flight Test Center History Office)

The X-15-3 on Rogers Dry Lake. (Air Force Flight Test Center History Office)

X-15 pilots pose with the X-15A-2 on December 2, 1965. From left to right; Air Force Captain Joseph H. Engle, Air Force Major Robert A. Rushworth, NASA pilot Jack McKay, Air Force Major William J. "Pete" Knight, NASA pilot Milton O. Thompson, and NASA pilot Bill Dana. (NASA Photo EC66-1017)

Northrop Lifting Bodies

Northrop designed a pair of rocket-powered, wingless lifting bodies, the M2-F2 and the HL-10, to evaluate the ability of low lift-to-drag vehicles to make precision landings after high-speed, unpowered descents from high altitudes through the use of energy management techniques. The M2-F2, N803NA was based on the design of the lightweight M2-F1, rounded on the bottom and flat on top with a pair of vertical stabilizers. The HL-10, N804NA was a new design that was flat on the bottom and rounded on the top with three vertical stabilizers. Northrop delivered the M2-F2 and the HL-10 to Edwards Air Force Base in 1965 and 1966, respectively. The lifting body program made its major contribution to America's space effort by validating the approach-and-landing techniques that would be employed by the space shuttle.

The Northrop lifting bodies were designed to utilize the air-launch capability provided by the Stratofortress motherships. Northrop Norair manufactured a twenty-two foot long steel frame adapter to attach either the M2-F2 or HL-10 to the X-15 pylon. The adapter supplied electrical power, fluids, and gases to the lifting body.

Wind tunnel drop tests of the NB-52 and the lifting bodies were conducted in the 7-foot by 10-foot high-speed tunnel at NASA Langley. Both lifting bodies displayed a tendency to roll to the right as they dropped from the pylon.

The process of mounting the lifting bodies to the X-15 pylon differed from that used for the X-15. The X-15 pylon was modified to accept a lift bar that extended from the sides of the front of the pylon. A block and tackle was attached to each end of the bar to haul the lifting body adapter up to the pylon on chains. Umbilical cables were connected to the adapter, and it was secured to four support hooks on the pylon. The lifting body was rolled into a pit near the NASA hangar that provided clearance under the adapter. The lifting body was then hoisted and attached to the adapter. Before flights the lifting body was serviced with breathing oxygen, hydrogen peroxide for the landing rockets, and nitrogen and helium pressurization gas. [54]

While mounted on the X-15 pylon, the lifting body cockpit was in front of the leading edge of the wing, allowing the pilot to eject in the event of an emergency. [55]

The M2-F2 was first carried aloft under the wing of the NB-

Northrop M2-F2 hangs from the Northrop Norair adapter on the pylon of the NB-52B on September 23, 1965. (NASA photo ECN873)

Northrop M2-F2 hangs from the Northrop Norair adapter on the pylon of the NB-52B on September 23, 1965. (NASA photo ECN882)

Northrop M2-F2 hangs from the Northrop Norair adapter on the pylon of the NB-52B on September 23, 1965. (NASA photo E13868)

Northrop M2-F2 hangs from the Northrop Norair adapter on the pylon of the NB-52B on September 23, 1965. (NASA photo E13865)

52B on two unmanned captive carry flights on October 21, 1965. Lt. Col. Fulton and Captain Charles Doryland took off and landed straight ahead on the lakebed runway. [56]

Each lifting body design was tested in a series of glide flights. The NB-52B climbed to an altitude of 45,000 feet and adopted a heading toward the northeast along a line east of Rogers Dry Lake. Following launch, the lifting body descended at a rate of 10,000 to 15,000 feet per minute for about four minutes. The pilot of the lifting body steered the vehicle through a 180-degree left turn onto the approach to Runway 17 or Runway 18 on Rogers Dry Lake. This took the lifting body a short distance north of Highway 58. The pilot aimed at the landing target on and initiated the flare at an altitude of about 1,500 feet and a speed of about 300 knots. After the flare, the airspeed of the lifting body decreased rapidly. The pilot deployed the landing gear several seconds before touchdown at a speed of about 200 knots. As the landing gear extended, the pilot pulled back on the stick to counteract a tendency for the nose

Northrop M2-F2 hangs from the Northrop Norair adapter on the pylon of the NB-52B on September 23, 1965. (NASA photo ECN873)

of the lifting body to pitch downward. There was no nose-wheel steering, so the pilot steered the lifting body during rollout by differential braking of the main gear.

Jerry Gentry described the landing procedure as a hectic affair. "You are always pretty busy because you were doing something. It was like kind of a dance routine that you went through. You practiced on the simulator what you were going to do at various points. You were looking at longitudinal stability, lateral/directional stability or whatever. You were busy flying a planned profile, but you still had time to look around and check the chase aircraft once in awhile. Once you got to base leg, you were pretty preoccupied with looking at where you were going to put the thing on the lakebed". [57]

Lifting Body Pilot's Ride, Under the Wing of the NB-52B

Lifting body pilots agreed that the experience of riding under the wing of the NB-52B was the least pleasant aspect of most flights. Jerry Gentry explained, "You spend a helluva lot of time hooked up to that B-52 because you got in the lifting body about 45 minutes prior to takeoff and then you taxi out and take off. It's 45 minutes to climb to 45,000 feet ... there was the sensation of helplessness when the B-52 is taking off. You get a great glimpse that you only ever get when you are flying the airplane single seat, but now you have absolutely no control over it ... I said, 'Well, I know how a bomb must feel if a bomb had feelings!' The worst thing though, worse than the takeoff, is when you had an abort and then you had to come back [under the B-52's wing] and go through the landing". [58]

After launching the sixty-second flight of the X-15-1 on November 4, 1965, the NB-52B underwent over half a year of maintenance and modification. On January 27, 1966, Major Bock and Major Gayland Jones flew the NB-52B to Tinker Air Force Base, Oklahoma for IRAN. [59] Major Bock and Major Doryland flew the NB-52B back to Edwards Air Force Base on April 8, 1966. Following its return from Tinker Air Force Base the NB-52B underwent five weeks of pylon modification to accept the increased weight of the X-15A-2 with its external tanks. [60]

Milton Thompson occupied the cockpit of the M2-F2 on its first manned captive-carry mission on June 7, 1966. Lt. Col. Fulton and Major Doryland piloted the NB-52B.

The first captive carry flight of the X-15A-2 with full external tanks was made on June 27, 1966. It was the heaviest payload carried by the NB-52B up to that date. The primary objective of the mission was a test of the propellant jettison system. Lt. Col. Fulton and Major Doryland were at the controls of the NB-52B.

NB-52B carries the X-15A-2 with full external fuel tanks on a captive-carry flight in on June 27, 1966. Lt. Col. Fulton and Major Charles Doryland piloted the Stratofortress. Col. Robert Rushworth occupied the cockpit of the X-15A-2. Martin MA-25S thermal protection ablative material has been applied to the ventral stabilizer and skids of the X-15A-2. Ice coats the Liquid Oxygen tank on the right. (AFFTC/HO)

The propelants are jettisoned from the X-15A-2 later on June 27, 1966. (AFFTC/HO)

Col. Robert Rushworth occupied the cockpit of the X-15A-2. The right outrigger landing gear of the Stratofortress stuck in the down position, but did not interfere with the flight.

Lt. Col. Fulton and Captain Jerry Bowline piloted the NB-52B as it carried the M2-F2 on the second manned, captive-carry flight on July 6. [61] Milton Thompson occupied the cockpit of the M2-F2.

The X-15A-2 was launched with full external tanks for the first time from the NB-52B over Mud Dry Lake on July 1, 1966. Lt. Col. Fulton and Major Doryland were at the controls of the Stratofortress. After launch, Col. Rushworth received an indication that the ammonia was not flowing from the external fuel tank. The draining of liquid oxygen from the other tank produced a rapid shift in the center of gravity to the right. Rushworth had to abort the flight. He shut down the XLR-99 engine, jettisoned the partially full fuel tanks and successfully glided the X-15A-2 to a landing. [62] It was Col. Rushworth's last X-15 flight.

First Heavyweight Lifting Body Flight

The first glide flight of the M2-F2 was scheduled for July 11, 1966. Lt. Col. Fulton and Major Doryland were taxiing the NB-52B out that morning when a Lockheed F-104 Starfighter blew a tire while landing on the main runway. Then a pre-launch check revealed that the hydrogen peroxide fueled landing rocket was not working. The launch was rescheduled for the next day.

Lt. Col. Fulton and Captain Bowline launched Milton Thompson on the first heavyweight lifting body flight on July 12, 1966. Victor Horton manned the launch panel operator's station. Thompson echoed Gentry's opinion about the helpless feeling of riding under the wing of the NB-52B. Thompson noted that the slowest part of the flight "was the 45 minutes I sat in the M2-F2 at the end of the pylon waiting for the B-52 to reach launch altitude." [63]

Jack McKay Lands at Smith Lake

Jack McKay was faced with his third X-15 emergency landing on September 8, 1966. Colonel Joseph Cotton and Major

X-15A-2 sits on Mud Dry Lake after Col. Rushworth's emergency landing on July 1, 1966. (NASA photo E15263)

Support aircraft on Mud Dry Lake following the aborted launch of the X-15A-2 on July 1, 1966. During each X-15 flight, a C-130 Hercules flew a racetrack pattern near the middle of the X-15's flight path. The C-130 immediately flew an emergency team to the site of any off base landing. The C-47 brought NASA personnel from the Flight Research Center to the landing site. (NASA Photo ECN1367)

A gathering of X-15 pilots and former X-15 pilots. From left to right; North American pilot Scott Crossfield, Air Force Captain Joe Engle, Air Force Major Bob White, NASA pilot Milt Thompson, Navy Commander Bob Rushworth, NASA pilot Jack McKay, and NASA pilot Bill Dana (North American Rockwell)

Hangar 4802 at the NASA Flight Research Center on August 30, 1966. All three X-15s are parked on the right. On the left, the aircraft are the HL-10, M2-F2, M2-F1, F-4A, F5D-1, F-104, and C-47. (NASA Photo EC66-1461)

The NB-52B takes off with the M2-F2 for its eighth glide flight on September 22, 1966. Colonel Cotton and Major Reschke piloted the Stratofortress. Bruce Peterson flew the M2-F2. The NB-52B has had its white nose radome replaced with a black radome. (NASA photo E15788)

The NB-52B carries the X-15A-2 before its fiftieth flight on November 18, 1966. Col. Joseph Cotton and Fitzhugh Fulton piloted the Stratofortress. Pete Knight flew the X-15A-2 to a speed of Mach 6.33 (4,261miles per hour) and an altitude of 98,900 feet. (AFFTC/HO)

The NB-52B carries the X-15A-2 before its fiftieth flight on November 18, 1966. (NASA Photo ECN1519)

HL-10 mated to the NB-52B in late 1966. (NASA Photo ECN1587)

HL-10 mated to the NB-52B on November 12, 1966. (NASA Photo)

T Doryland piloted the NB-52B. Jack Russell manned the launch panel operator's station. Shortly after the X-15-3 was launched over Smith Dry Lake, McKay saw that the fuel pressure was too low. At the recommendation of the ground controller, McKay throttled back to 50%, but the fuel pressure remained too low. McKay shut down the XLR-99 rocket engine and turned the X-15-3 around. He successfully jettisoned the propellants and landed on the dry lakebed. It was his last flight in an X-15.

NB-52 Utilization

In the six months of May to October 1966, flights of the X-15 required seventeen percent and six percent of the available working days of the NB-52A and NB-52B, respectively. Forty-seven percent of the available working days were devoted to NASA preflight and post-flight tests, inspection, and maintenance. [64]

Colonel Cotton and Major Sturmthal flew the NB-52B past the crowd at the Pt. Mugu Naval Air Station airshow on October 29 and October 30, 1966.

The Northrop HL-10 was first mated to the NB-52B on November 12, 1966. The first captive-carry flight of the HL-10 was recorded in the flight log but never flown. The HL-10 was de-mated from the NB-52B, and the X-15A-2 was mounted on the pylon for its fiftieth flight.

An attempt was made to launch two X-15s on November 18, 1966. Plans called for Major Knight to make an attempt at a new speed record in the X-15A-2 and for Bill Dana to make a flight in the X-15-3. The X-15A-2 was mated to the NB-52B, and the X-15-3 was mated to the NB-52A.

Fitzhugh Fulton and Colonel Cotton took the NB-52A aloft with the X-15-3. Jack Russell manned the launch panel operator's station. The Hadley transformer aboard the X-15-3 malfunctioned and the flight was aborted.

Fulton and Cotton landed the NB-52A with the X-15-3 still mounted on the pylon. They climbed into the NB-52B and took off with Pete Knight in the X-15A-2. It was the second flight of the X-15A-2 with full external fuel tanks. The X-15A-2 was launched over Mud Lake. Knight accelerated to Mach-6.33 (4,250 miles per hour), a new speed record. It was the first time that the X-15 had exceeded Mach-6 since November 9, 1961.

The NB-52B carried the HL-10 on its first captive-carry flight on December 20, 1966. Major Reschke and Major Sturmthal piloted the Stratofortress. Bruce Peterson occupied the cockpit of the HL-10. The collection of test data was so successful that second planned captive-carry flight was cancelled.

Captain Jerauld Gentry pilots the M2-F2 to a landing at the conclusion of its fourteenth glide flight in the company of NASA Lockheed F-104N 812 on November 21, 1966. Col. Joseph Cotton and Fitzhugh Fulton piloted the NB-52B on that mission. (NASA photo EC66-1567)

Air Force Capt. Jerauld Gentry stands next to the M2-F2 following a glide flight in 1966. (NASA Photo EC97-44183-1)

First HL-10 Glide Flight

Col. Cotton and Major Reschke launched Bruce Peterson from the NB-52B in the HL-10 on its first glide flight on December 22, 1966. Immediately after launch the HL-10 started vibrating. The yaw axis of the stability augmentation system entered a limit rate cycle. It was ordering control surface movements at a rate that exceeded the ability of the actuators to move them, The pilot of the HL-10 had control of the gain of the channels of the stability augmentation system. Under the direction of the ground station, Peterson reduced the gain of the yaw channel, halting the limit rate cycle and stopping the vibration. The pitch channel of the stability augmentation system entered a limit rate cycle. The ground station instructed Peterson to reduce the gain of the pitch channel several times during the flight.

Peterson performed a practice flare at altitude to determine how the HL-10 behaved at low speed and high angle of attack. As the airspeed dropped off, the controls of the HL-10 became ineffective. Peterson pulled the control stick all the way back and to the left, but the lifting body did not respond. Control of the lifting body returned when Peterson dropped the nose and gained airspeed. He realized that the HL-10 would be uncontrollable at the planned landing speed of 200 knots.

Due to the many distractions during the flight, Peterson was late initiating the flare for landing. He pulled the nose up sharply and deployed the landing gear. The HL-10 was traveling at 280 knots when it touched down on the clay surface of the lakebed. It bounced and became airborne for several seconds before settling on the lakebed. [65]

Flight control engineer Wen Painter described the action from the point of view of those on the ground "Peterson said 'What's going on down there?' I said, 'Hell, it looks okay to me.' We didn't talk directly to the pilot. You could hear direct, but we never were allowed to talk to him. The controller and pilot talked back and forth. He [Peterson] was concerned about the control system but it looked okay to me from what I could see. From then on, everything got progressively worse and worse and worse ... This aircraft was essentially unstable for portions of the flight. We were changing gains and everything just kept getting worse. There were limit cycles because the airplane control surfaces were vibrating, moving at about four cycles. Everything was moving and shaking. We kept shutting the gains down lower and lower and only had one position to go when he turned final, and I knew it was going to get worse... I knew he was going to have to drop the landing gear. We had no idea what was going to happen when he dropped the landing gear. He may get into a PIO [pilot induced oscillation] close to the ground. We decided to leave it there and it stabilized. We had about plus or minus three or four degrees of control surface movement there. He could feel the vibration in his hands because of the frequency of the augmentation system driving when he landed. I said it was the most terrifying 300 seconds of my life!" [66]

The stability of the HL-10 was so poor that it had to be modified to improve its flight characteristics. It did not fly again until March 15, 1968.

NB-52B Pylon Modified for Lifting Body Adapter

Following the the first glide flight of the HL-10, the NB-52B made a series of Air Force flights. In February 1967, A pair of hooks were added to the X-15 pylon a little behind the forward X-15 attachment hook to provide four-point attachment for the lifting body adapters. [53]

Following the lifting-body glide flight tests, each lifting body had an XLR-11 rocket engine installed. These were the same engines that had been used for the initial flights of the X-15.

Lifting body flights were of shorter duration and lower speed than X-15 flights. They were conducted in the vicinity of

Bruce Peterson flares the Northrop HL-10 just before landing on December 22, 1966. (NASA photo E661598)

Edwards Air Force Base. Flights were launched no farther away than Cuddeback Dry Lake. Following launch, the lifting body pilot ignited two to four of the thrust chambers of the rocket engine. On a typical flight, he climbed under power for something over two minutes to an altitude of about 70,000 feet. The pilot shut off the engine after the planned engine burn time or allowed it to shutdown when the fuel was exhausted. The flight data relevant to lifting body reentry was collected primarily after engine shutdown. The approach and landing were essentially the same as for the glide flights. [67]

Jerry Gentry was scheduled to make the fifteenth flight of the M2-F2 on April 20, 1967. The XLR-11 rocket engine had just been installed in the M2-F2 in preparation for a series of powered flights. The M2-F2 was mated to the NB-52B, but the launch attempt was aborted. The M2-F2 was de-mated from the NB-52B, so that the X-15-3 could be mated for its fifty-eighth flight.

Bill Dana Lands on Silver Lake

Bill Dana made an emergency landing on Silver Lake in the X-15-3 after launch from the NB-52B on April 26, 1967. Colonel Cotton and Major Bowline piloted the Stratofortress. Jack Russell manned the launch panel operator's station. The ball nose malfunctioned during the pre-flight check. It was not critical to the mission, so the X-15-3 was launched anyway. A frozen sensing line produced a false low fuel-pressure reading, so Dana shut down the XLR-99 rocket engine 23 seconds after ignition. He turned the X-15 around and landed safely on the dry lakebed.

Jerauld Gentry piloted the fifteenth glide flight of the M2-F2 on May 2, 1967. Colonel Joseph Cotton and Major William Reschke piloted the Stratofortress. The NASA Flight Research Center Diary entry for the day reads, "Bird flew through wing of MotherShip 008". Perhaps the bird would have worded the description of the incident a little differently.

Dummy Ramjet Test

Pete Knight was scheduled to fly the X-15A-2 with the dummy ramjet on the ventral fin for the first time on May 5, 1967. The test shape was intended to examine the aerodynamic impact that an air-breathing ramjet would impose on the X-15. It had no internal airflow.

It was expected that emission of gasses from the ablative material that was to be used on faster flights of the X-15A-2 might obscure the pilot's view through the windscreen. To allow the pilot to see out of the cockpit for landing, an "eyelid" was attached to the cockpit canopy. The eyelid would cover the left windscreen during high-speed flight. At low speed it would be opened so that

The M2-F2 flares for landing on Rogers Dry Lake after launch from the NB-52B on May 2, 1967. (NASA Photo EC1707)

the pilot would have at least one clear windscreen. The eyelid was installed for the first time on this flight. It carried a test patch of pink ablative material.

Colonel Cotton and Major Reschke intended to launch the X-15A-2 over Hidden Hills Dry Lake, but the flight was aborted due to weather in the launch area. [68]

The test of the ramjet test shape and eyelid was conducted on May 8, 1967. Col. Cotton and Major Bowline launched Pete Knight in the X-15A-2 from the NB-52B. Knight achieved a top speed of Mach-4.75 (3,193 miles per hour). The X-15A-2 reached an altitude of 97,600 feet.

M2-F2 Landing Accident

Bruce Peterson was severely injured and the M2-F2 was nearly totaled in a landing accident during its sixteenth glide flight on May 10, 1967. Col. Cotton and Major Bowline launched the M2-F2 from the NB-52B. The first part of the flight proceeded normally. As Peterson turned to line up with the runway, he dropped the nose of the lifting body slightly to keep it pointed at the pre-flare aim point. At the reduced angle of attack, the M2-F2 became unstable about its roll axis and entered a "Dutch roll".

Peterson recovered from the Dutch roll by pulling the nose up, but the M2-F2 was no longer pointed at the runway marked out on the lakebed. A Piasecki CH-22 rescue helicopter was positioned

The NB-52B prepares to take off from Rogers Dry Lake with the X-15A-2 on an aborted attempt to launch its fifty-first flight on May 5, 1967. The X-15A-2 is carrying the ramjet test shape on its ventral fin for the first time. Col. Joseph Cotton and Major William Reschke piloted the Stratofortress. Jack Russell manned the launch panel. Pete Knight occupied the cockpit of the X-15A-2. The mission was aborted due to bad weather at the launch site. . (NASA Photo E16780)

Pete Knight accelerates away from the NB-52B in the X-15A-2 on its fifty-first flight on May 8, 1967. The X-15A-2 is flying with the ramjet test shape for the first time. The X-15A-2 reached a speed of Mach-4.75 (3,205 miles per hour) and an altitude of 97,600 feet. (Air Force Flight Test Center History Office)

The X-15A-2 glides back to Edwards Air Force Base at the conclusion of its fifty-first flight. (AFFTC/HO)

Pete Knight glides back to Rogers Dry Lake in the X-15A-2 after engine shut down during its fifty-first flight on May 8, 1967. Thermal protection ablative was applied to the clamshell eyelid over the left windshield. (NASA Photo EC67-1731)

to the east of the intended landing point, near the mid-point of the runway. Peterson became concerned about the possibility of colliding with the helicopter and called for it to move. The helicopter pilot turned and headed away from the area that the M2-F2 was headed for.

The lakebed where Peterson touched down lacked runway markings, making it difficult for him to estimate his height above the featureless surface. The M2-F2 touched down on the lakebed before Peterson hit the landing gear switch. The first impact marks were made by the rounded bottom of the vehicle. The telemetry antenna on the bottom of the lifting body was destroyed on impact, and radio telemetry ceased immediately.

The landing gear of the M2-F2 started to extend as the lifting body bounced off the lakebed. The opening landing gear doors made the next set of marks eighty feet farther along the lakebed. Then the landing gear struts were ripped away or bent over. The flight ended in a spectacular rolling, flipping crash, film of which was used in the opening sequence of *The Six Million Dollar Man* television series. Peterson was badly injured in the accident and subsequently lost his sight in one eye due to his injuries.

Bruce Peterson recalled the event, "On that particular day we'd had a wind pickup and as I pushed over to low angle of attack, I saw the nose move out to the right and I thought I'd hit a wind-shear. I used some rudder to bring the nose back and it came back and I got a roll associated with it and I countered the roll and immediately was in a pilot-induced oscillation, extremely high roll oscillations, to the point I was banging my head on the canopy, quite hard. I knew I had to increase the angle of attack to get out of the maneuver. I had to do it so I would come out with wings level, so I kind of timed it and jerked on it when I thought it would recover wings level and it did. At this time though, my heading was pulled off the main runway and I didn't have enough altitude or room to maneuver back to the runway I was aiming for. There was a helicopter in front of me that surprised me. It was where I thought I was going to land. I told him to move west. I flared for the landing, and I flared to get underneath him or just behind him so as not to take us all out. Either I hit the rotor wash or I misjudged my altitude. I lit the emergency rocket first, to give myself more time. Then I put the gear down and before the gear got all the way down, one gear door caught on the lakebed and that flipped the vehicle. It tumbled, I guess about six times. Kind of the end of the game that day." [69]

Sixteen glide flights of the original M2-F2 configuration were flown from July 12, 1966 to May 10, 1967. The NB-52B launched the M2-F2 twelve times. Following Bruce Peterson's accident, the M2-F2 was rebuilt to fly again in a modified configuration.

The NB-52B was displayed at the Edwards Air Force Base Open house on May 21, 1967 with the X-15-1 sitting on the ground under the right wing and the Northrop HL-10 sitting under the left wing. Five days after the Open House, the NB-52B launched the X-15-3 on its sixtieth flight on June 22. Lt. Colonel Sturmthal and Colonel Cotton launched Bill Dana on a flight that reached a speed of Mach-5.44 (3,611 miles per hour) and an altitude of 82,200 feet.

Northrop M2-F2 lies on its back on Rogers Dry Lake following Bruce Peterson's landing accident on May 10, 1967. (NASA photo E16730)

Northrop M2-F2 lies on its back on Rogers Dry Lake following Bruce Peterson's landing accident on May 10, 1967. (NASA photo E16738)

Bill Dana stands in front of the X-15-3 following a flight in 1967. (NASA Photo ECN-1716)

Complete Electrical Failure of the X-15-1

Pete Knight was faced with a complete electrical failure of the X-15-1 on June 29, 1967. Major Reschke and Major Bowline piloted the NB-52B. Jack Russell manned the launch panel operator's station. The X-15-1 was launched over Smith Ranch Lake. Knight ignited the XLR-99 engine and pulled back on the stick to begin the climb to 250,000 feet. Sixty-nine seconds after ignition, just as NASA-1 advised Knight that he was passing through 104,000 feet, the XLR-99 engine shut down. The inertial system computer gross malfunction light came on, followed by the three stability augmentation system warning lights and both generator warning lights. Then all the lights in the cockpit went dark. The ground station stopped receiving telemetry transmissions from the X-15-1.

The NB-52B was displayed at the May 21, 1967 Edwards AFB Open House with the X-15-1 sitting on the ground under the pylon on the right wing. The Northrop HL-10 was on static display on the left side the NB-52B. (Paul Minert via Greg Spahr)(NASA photo E16738)

As both auxiliary power units stopped working, Knight lost his electrical and hydraulic power. He lost all control of the aerodynamic control surfaces as he ascended to over 170,000 feet, and his reaction control thrusters did not produce the desired attitude adjustments.

Knight managed to restart one of the auxiliary power units. Although the mission time line called for a landing on Grapevine Lake after 69 seconds, Knight decided to turn back to Mud Lake, which is much larger and easier to find. Knight had to fly the reentry by the seat of his pants, without the aid of attitude instruments. He pulled back on the stick to raise the angle of attack until he felt the X-15-1 become unstable in yaw. He lowered the nose and raised it again until he felt that he could maintain stable flight while pulling out of he dive. Knight rolled the X-15-1 into a 6-g left-hand turn toward Mud Lake.

For over eight minutes, the X-15-1 was lost to the ground station and the chase planes in the air. It was only as it crossed the edge of Mud Lake that it was spotted by Chase 2.

The flaps did not deploy during the landing, so the X-15-1 touched down at over 250 knots. The skid under the ventral fin failed to extend. Despite the resulting high loads on the landing gear struts, Knight made a successful landing on the dry lakebed, sliding to a stop after 9,050 feet.

After the emergency landing, Knight had difficulty getting the cockpit canopy raised without assistance from the outside. It took several pulls on the release handle to get it to open. Then Knight was unable to release another connection that held the canopy down. He removed his helmet to relieve the buildup of heat inside his suit. He finally resorted to pulling the emergency canopy release. The seat headrest blew off, hit the canopy, cracking the inner pane of the left windshield, and came back down on Knight's head. [70]

In early July 1967, a liquid oxygen filter modification required for upcoming X-15A-2 flights was performed on the X-15 pylons of both NB-52s. [71]

X-15A-2 Turns White

In order for the X-15A-2 to survive the higher temperatures generated by flight at speeds above Mach-7, the entire spaceplane was coated with Martin MA-25S spray-on, ablative thermal protection material. The pink ablative material was covered by a white sealant.

It took three attempts to launch Pete Knight in the X-15A-2 on its fifty-second flight in August 1967. The entire spaceplane was coated with ablative thermal protection material for the first time, but was not carrying the external fuel tanks. A circular-inlet supersonic combustion ramjet test shape was mounted on the ventral fin. Colonel Joe Cotton and Major Reschke flew the NB-52B on each launch attempt.

The first launch attempt, scheduled for August 14, was postponed until August 16 due to weather. The nine o'clock take-off for the second launch attempt was delayed until ten o'clock and then ten-fifty. Just before noon, the number 2 auxiliary power unit failed, and the flight was aborted again.

Col. Joe Cotton and Major Reschke launched Pete Knight in the X-15A-2 over Hidden Hills Lake on August 21, 1967. The X-15A-2 reached a speed of Mach-4.94 (3,368 miles per hour) and an altitude of 91,000 feet.

The nB-52B takes off with the X-15A-2 in the all-up configuration on August 7, 1967. (AFFTC/HO)

The NB-52B carries the X-15A-2 in the all-up configuration on a captive-carry flight on August 7, 1967. (AFFTC/HO)

The X-15A-2 jettisons propellants during a captive carry flight on August 7, 1967. Ice no longer coats the liquid oxygen tank. (AFFTC/HO)

The NB-52B carries the X-15A-2 folowing the propellant jettison test on August 7, 1967. (AFFTC/HO)

The NB-52B carries the X-15A-2 in the all-up configuration on a captive-carry flight on August 7, 1967. (U.S. Air Force Museum)

Martin-Marietta X-24A

The third heavyweight lifting body, the Martin-Marietta X-24A, 66-13551 was delivered to Edwards Air Force Base by C-130E, 63-7843 on September 14, 1967. The design of the X-24A was based on the shape of the SV-5. The X-23 was a sub-scale SV-5 that was lofted into sub-orbital flight by a rocket. It demonstrated the behavior of the shape during flight at high Mach numbers. The rocket powered X-24A was intended to evaluate its flight behavior during the transition from supersonic to subsonic flight and at landing. A jet powered version, the SV-5J was intended to take off and land conventionally, but was never flown.

As had been done for the Northrop lifting bodies, wind tunnel

The NB-52B carries the X-15A-2 in the white ablative coating without external tanks in August 1967. (AFFTC/HO)

Pete Knight accelerates away from the NB-52B in the X-15A-2 in the white ablative coating without external tanks on August 21, 1967. (AFFTC/HO)

drop tests of the NB-52 and X-24A were conducted in the 7x10 foot high-speed tunnel at NASA Langley. The tests determined the proper carry angle for the pylon adapter. In the wind tunnel tests, the X-24A displayed the same tendency to make an abrupt but controllable roll to the right after launch that the Northrop lifting bodies had. [72]

Foot Operated Intercom Switch

Eight years into the X-15 program a foot-operated intercom switch was installed at the launch panel operator's station in September 1967. The switch installation took sixteen man-hours and cost eighty dollars. [73]

The X-15A-2 in the "all-up" configuration hangs from the wing of the NB-52B in the background in October 1967. The X-15-3 hangs from the pylon of the NB-52A in the foreground. (AFFTC/HO)

All-Up X-15 Flight

The NB-52B launched the X-15A-2 on its only flight in the "all-up" configuration on October 3, 1967. It carried the ablative coating, external tanks, and the ramjet test shape. Col. Cotton and Major Reschke launched Pete Knight from the NB-52B in the X-15A-2 over Mud Lake in Nevada at an altitude of 45,000 feet. It flew nearly directly over Racetrack Playa in Death Valley National Monument at about 100,000 feet going over four thousand miles per hour. The X-15A-2 achieved Mach-6.70 (4520 miles per hour), the highest speed attained by any manned aircraft to that time. Only the Space Shuttles have flown faster,

As the X-15A-2 crossed Death Valley, shock waves streaming from the dummy ramjet mounted on the ventral stabilizer superimposed themselves in unexpected ways, creating hot spots that burned holes into the nickel-iron skin of the ventral fin. Superheated air flowed into the stabilizer and the rear fuselage of the X-15A-2, burning through the plumbing and wiring within.

Three of the four explosive bolts holding the dummy ramjet cooked off. The single remaining bolt eventually failed and the dummy ramjet fell off of the X-15A-2 as it approached the lakebed. The ramjet was recovered after analysis of the flight data showed a decrease in the rate of deceleration that resulted from the reduction in drag as it fell off.

After landing, Pete Knight was taken by surprise when the recovery team all headed to the back of the X-15A-2, instead of helping him to get out of the cockpit. Their attention was grabbed by the severely burned appearance of the aft fuselage.

The damage to the X-15A-2 was subsequently repaired, but the next year funding for the program was discontinued. The X-15A-2 never flew again. The first supersonic combustion ramjet flight had to wait for the X-43A Hyper-X program to be flown twenty-seven years later. The X-15A-2 is now displayed at the National Museum of the Air Force.

Major Knight Qualifies for Astronaut's Wings

Major Pete Knight became the fourth X-15 pilot to qualify for astronaut's wings following launch from the NB-52B on October 17, 1967. Major Reschke and Squadron Leader John Miller piloted the Stratofortress. They launched the X-15-1 over Smith Ranch Lake. The X-15-1 was going Mach 5.45 when the XLR-99 engine burned out 84.2 seconds after ignition. Knight reached an altitude of 280,500 feet. The maximum Mach number of 5.53 (3,856 miles per hour) occurred during the descent from the peak altitude.

The NB-52B launches Pete Knight in the X-15A-2 on its fifty-third flight on October 3, 1967. (U.S. Air Force Museum)

The X-15A-2 drops away from the NB-52B on October 3, 1967. It was the fastest flight of the program and the last flight of the X-15A-2. Pete Knight piloted the X-15A-2 to a speed of Mach-6.70 (4,520 miles per hour).(AFFTC/HO)

The scorched X-15A-2 following the all-up flight in October 1967. (NASA photo EC1834)

Loss of Mike Adams and the X-15-3

The X-15-3 was lost and its pilot, Mike Adams, was killed following launch from the NB-52B on November 15, 1967. It was Adams' seventh flight in the X-15 and the one hundred ninety-first flight of the program. The third X-15 was equipped with the Minneapolis-Honeywell MH-96 adaptive-gain flight control system that automatically blended the aerodynamic flight controls with the reaction control thrusters as the X-15 transitioned from the atmosphere to the vacuum of space and then re-entered the atmosphere.

The sixty-fifth flight of the X-15-3 was scheduled for October 31. Fifteen seconds before launch, a malfunction of the XLR-99 rocket engine resulted in a launch abort. The mission was postponed for two weeks.

Col. Joe Cotton and Squadron Leader Miller launched Mike Adams from the NB-52B in the X-15-3 at 10:30 in the morning over Delamar Dry Lake on November 15.

The X-15 carried seven experiments; a boost guidance experiment, a solar spectrum measurement, an ultraviolet rocket exhaust plume detector, a micrometeorite collector, a tip pod deflection camera, samples of Saturn rocket booster ablative thermal protection material, and a "traversing probe" to measure the position of the shock wave in front of the wing tip pod. The traversing probe was an air pressure sensor mounted in the nose of the wing tip pod. It moved back and forth to detect the position of the shock wave relative to the probe tip.

These experiments added to the workload of the X-15 pilot. The solar spectrum measurement required the pilot to extend it from the q-bay behind the cockpit and then retract it later in the flight. The ultraviolet plume detector required the X-15 to be rolled from side-to-side so that the detector could observe the X-15 rocket exhaust plume against black space and then against the earth as a background.

Adams accelerated for 82 seconds to a speed of Mach-5.20. The X-15-3 ascended on a ballistic trajectory to 266,000 feet. By exceeding an altitude of fifty miles, Adams earned his U. S. Air Force astronaut wings.

The traversing probe experiment was not properly qualified for flight in the vacuum above 100,000 feet. It was not realized that a motor used in the experiment depended on air for electrical insulation. In a vacuum, electrical discharges arced between the components of the motor. As the X-15 reached high altitude, the arcing inside the motor produced an electrical disturbance that propogated through the X-15.

During the ascent, the electrical disturbance from the traversing probe experiment caused the automatic reaction control thrusters to shut down. It also caused errors in the presentation of the attitude of the X-15 on the cockpit displays. The thrusters

Mike Adams. (North American Rockwell)

should have automatically kept the X-15 aligned with its flight path. The failure of the thrusters allowed the nose of the spaceplane to swing to one side as it reached its peak altitude.

Twice during the ascent, the inertial computer system failure lights came on. Then an electrical disturbance disengaged the pitch and roll stability dampers. Five seconds later the dampers re-engaged.

As the X-15-3 started to descend, Adams switched to the manual reaction controls. It is believed that he was suffering from vertigo. He made control inputs in the wrong direction, causing the X-15 to yaw farther from the direction of flight. The X-15 spun completely around at 3,500 miles per hour as it re-entered the atmosphere over Death Valley.

As the X-15-3 descended through 120,000 feet, Adams recovered from the spin, but at that point the adaptive flight control system drove the elevators into a rate-limit cycle. The X-15 began a series of violent pitch and roll oscillations. Mike Adams was tossed up and down inside the cockpit of the X-15 with thirteen times the force of gravity. He was unable to initiate the ejection sequence.

The g-load on the X-15 caused its nickel-steel structure to break into many pieces at an altitude of about 60,000 feet over the Panamint Mountains heading southwest toward Edwards Air Force Base. It had rolled onto its left side when the upper vertical stabilizer failed. The stabilizer pitched over, causing even greater side loads that overloaded the welds between the oxygen and fuel tanks. The fuselage broke in half. The two halves pulled apart, tearing piping, supports, and wiring out of the tunnels on the fuselage sides. These smaller components decelerated very rapidly and drifted with the wind to the north of the flight path.

Almost instantly the wings and stabilizers separated from the rear fuselage. They were torn into pieces of sheet metal and structural members, which carried another three to five miles.

The rear fuselage tank section hurtled a little farther. The engine ripped out of the rear fuselage and sailed farther west along with the relatively dense wingtip pods.

The forward fuselage was the heaviest, densest, and aerodynamically cleanest piece of the wreckage, so it traveled the greatest distance. As it passed over the Lava Hills, the cockpit canopy peeled off, striking Mike Adams' helmet and leaving black

The wreckage of the cockpit of the X-15-3 lies in the desert after Mike Adams fatal crash on November 15, 1967. (NASA photo E17879)

The wreckage of the cockpit of the X-15-3 lies in the desert after Mike Adams fatal crash on November 15, 1967. (NASA photo E17873)

The wreckage of the rear fuselage of the X-15-3 lies in the desert after Mike Adams fatal crash on November 15, 1967. (NASA photo E18073)

paint marks on it. The cockpit camera tore from its mount in the canopy, and the film cartridge fell from the camera. The film came down above the rest of the recovered debris on the eastern side of the mountains. A 16-inch high-pressure helium sphere landed in that same area. The camera and the canopy landed just short of the rocket engine.

The forward fuselage, carrying the unconscious Adams, plummeted to the ground west of the Trona Road. It landed on its right side, flattening the fuselage and driving some of the structural elements through the right side fuselage skin. It bounced and rolled over, coming to rest upside down.

Mike Adams was killed when the cockpit section hit the ground near the town of Johannesburg in the Mojave Desert five minutes and fifty seconds after he had been launched from the NB-52B. He was posthumously awarded astronaut's wings.

NB-52B Engine Change

The NB-52B was out of service in early December 1967 while it underwent an engine change. It then was subjected to a phase inspection, which kept it out of service until mid-January 1968.

Breathing Oxygen Pressure Gauge Installation

The investigators of Mike Adams' crash criticized the lack of gauges showing the oxygen source pressure and the regulated X-15 pilot breathing oxygen pressure. Two oxygen pressure gauges were installed on top of the cable tunnel on the right hand side of the NB-52B in the lower cabin. The gauges showed the pressure of the breathing oxygen being supplied to the X-15 and the pressure of the supply of breathing oxygen aboard the NB-52B. The installation of the gauges took twenty-nine man-hours and cost $161.00. [74]

Second Flight of the HL-10

Preparations for the second glide flight of the HL-10 began in early 1968. A taxi test of the HL-10 was performed on February 12. The results were "not good".

HL-10 flights resumed with a manned captive-carry flight on the NB-52B, piloted by Lt. Col. William Reschke and Squadron Leader Miller, on February 29, 1968. Captain Jerauld Gentry occupied the cockpit of the HL-10.

The second glide flight of the HL-10 was scheduled for March 12, 1968. High winds at Edwards Air Force Base resulted in a cancellation of the mission on that day and on March 14. Lt. Col. Reschke and Col. Cotton launched Captain Gentry from the NB-52B in the HL-10 on its second flight on March 15.

Lt. Colonel Sturmthal and Lt. Colonel Reschke flew the NB-52B past the crowd at the Edwards Air Force Base open house on May 19, 1968.

The NB-52B launched an X-15 for the last time on June 12, 1968. Lt. Col. Reschke and Col. Cotton launched Bill Dana in the X-15-1 on its seventy-seventh flight. Jack Russell manned the launch panel operator's station. Dana reached a speed of Mach-5.15 (3,563 miles per hour) and an altitude of 220,100 feet.

The NB-52B launched the X-15-1 thirty-seven times and carried it on twenty launch aborts and one captive-carry flight. The Stratofortress launched the X-15-2 twenty times in its original configuration and carried it on eight launch aborts. It launched the X-15A-2 fifteen times and carried it on eight launch aborts and six captive-carry flights. It launched the X-15-3 thirty four times and carried it on ten launch aborts and one captive-carry flight. The Stratofortress carried the three X-15s a total of 160 times and launched them 106 times.

Bill Dana piloted the last X-15 flight on October 24, 1968. He was launched from the NB-52A, 52-0003 in the X-15-1. The X-15-1 is displayed in the Milestones of Flight Gallery of the National Air and Space Museum.

The HL-10 hangs from the adapter on the pylon of the NB-52B on February 29, 1968. (NASA Photo E-18519)

The NB-52B carries the HL-10 to launch altitude in the company of a NASA Douglas F5D Skylancer chase plane. (AFFTC/HO)

Bill Dana stands in front of the HL-10 after its 24th flight on September 3, 1969. (NASA Photo ECN-2297)

First Powered Flight of the HL-10

The NB-52B carried the HL-10 aloft on a captive-carry flight on August 22, 1968. The take-off was delayed by problems encountered during the loading of the HL-10's liquid oxygen tank. Lt. Colonel Reschke and Squadron Leader John Miller piloted the Stratofortress. Victor Horton manned the launch panel operator's station. In the HL-10, John Manke checked out the operation of the XLR-11 rocket engine priming and fuel jettison.

Captain Jerauld Gentry checked out the operation of the XLR-11 during another HL-10 captive-carry flight under the wing of the NB-52B on August 30. The same crew was aboard the Stratofortress as on the previous flight. An attempt to launch the HL-10 from the NB-52B on September 4 was "aborted due to no pitch backup".

The first powered HL-10 flight was launched from the NB-52A. An attempt was made to launch the HL-10 on its first powered flight on October 22, 1968. The source pressure for the fuel and liquid oxygen tanks fell to a value that required scrubbing the mission.

Jerry Gentry was launched from the NB-52A in the HL-10 on October 23, 1968. Lt. Colonel Sturmthal and Squadron Leader Miller piloted the NB-52A. Following launch, one of the thrust chambers of the XLR-11 engine failed to light. Gentry attempted to start the engine three times and then shut it down. He jettisoned the propellants and headed for a landing on Rosamond Dry Lake. As he descended through 25,000 feet altitude, the stability augmentation system malfunctioned and put the HL-10 through a series of oscillations. Gentry reduced the gain setting to the flight controls, regained control of the lifting body and landed safely. The HL-10 was far from its intended landing site, so it took a long time for the emergency vehicles to reach Gentry. [75]

Gentry described the event: "When I launched, I couldn't get the rocket lit. I ended up going to Rosamond Lake, which was the only place to go out west. I jettisoned the fuel and put it on Rosamond and waited for 20 minutes for somebody to come and get me out of the thing." [76]

The NB-52B launched the first successful powered flight of a lifting body on November 13, 1968. Fitzhugh Fulton and Captain Edward McDowell launched John Manke in the HL-10 on its thirteenth flight. Victor Horton manned the launch panel operator's station. Manke fired two of the XLR-11 engine's thrust chambers for 186 seconds. The HL-10 reached a speed of Mach 0.84 (524 miles per hour) and an altitude of 42,650 feet.

UHF Radio

Communication between the NB-52B and the NASA ground station was through UHF radio. The NB-52s were equipped with primary and alternate ARC27 UHF radios. The radio had a remote head at the pilot's station, which allowed the pilot to select radio channels. The launch panel operator controlled the frequencies to which the channels were tuned. The mechanism for tuning the radio was hard to see from the launch panel operator's station. Sometimes the lock nuts holding the preset frequencies worked loose. The preset frequencies would be lost and the communication frequencies had to be tuned in manually. Tuning the radio manually resulted in long periods of time when the NB-52B was out of contact with the ground.

Adding to the problem was the fact that the primary radio was mounted in the tail of the NB-52B where it was inaccessible. If it failed, the launch panel operator had to switch to the alternate

The NB-52B carries HL-10 before its twenty-ninth flight on November 17, 1969. Fitzhugh Fulton and Captain Nicholas Fritz piloted the Stratofortress. Victor Horton manned the launch panel. Bill Dana piloted the HL-10 to a speed of Mach-1.59 (1,052 miles per hour) and an altitude of 64,590 feet (NASA photo EC69-2351)

The X-24A hangs from its adapter on the pylon of the NB-52B before its second flight on May 8, 1969. Fitzhugh Fulton and RAF Squadron Leader John Miller piloted the Stratofortress. Victor Horton manned the launch panel. Captain Gentry . The X-24A used a new pylon adapter that suspended it higher off the ground than the Northrop Norair adapter. (NASA photo EC69-2196)

radio, which was located behind his seat and considered to be nearly impossible to tune. To change the frequencies of the alternate radio manually, the launch panel operator had to undo his set belts, rise from his ejection seat, take his parachute off, disconnect from the oxygen source, turn around and peer through the ejection seat structure.

In August of 1968 a control head with frequency controls was requested for each UHF radio at the pilot's station. The pilot then had complete control of the communication equipment. The replacement parts were estimated to cost $778.00 and require 250 man-hours to install. [77]

The modification order was signed on December 13, 1968, but the installation did not take place until the following June. The final cost was $4,074.26 and 439 man-hours were expended over ten days. [78]

Major Edward McDowell and Captain George Larson flew the NB-52B to Tinker Air Force Base, Oklahoma for periodic maintenance on December 2, 1968. Captain Floyd Stroup and Squadron Leader Miller flew the Stratofortress back to Edwards Air Force Base on March 12, 1969.

First X-24A Captive Carry Flight

The NB-52B carried the Martin-Marietta X-24A on its first captive-carry flight on April 4, 1969. Lt. Col. Sturmthal and Major Jerry Bowline piloted the Stratofortress. Captain Jerauld Gentry occupied the cockpit of the X-24A. Victor Horton manned the launch panel operator's station.

The first glide flight of the X-24A occurred on April 17, 1969. Lt. Colonel Sturmthal and Captain Stroup launched Jerry Gentry from the NB-52B. Lt. Colonel Sturmthal and Captain Stroup then climbed into the NB-52A to launch John Manke in the HL-10 on its fifteenth flight on that same day.

Rebuilding the M2-F2 into the M2-F3

The severely damaged M2-F2 was reinstalled in the manufacturing jigs at Northrop. Skins and structural components were repaired and replaced. Eventually NASA Headquarters gave approval to fund the complete rebuilding of the lifting body. $700,000 was spent on the effort to rework the M2-F2 into the M2-F3. The most obvious difference was the addition of a central fin that acted as a splitter. Its primary function was to prevent the nose from swinging to the side when the upper flaps were moved in opposite directions to roll the lifting body rather than to add

The NB-52B takes off from Runway 04 with the X-24A. (U.S. Air Force Museum)

directional stability as the central fin of the HL-10 was intended to do. [79]

The controls of the M2-F3 were rigged to allow both of the rudders on the outer fins to be deflected outward at the same time to act as a speed brake. This was a feature that the M2-F2 and HL-10 had lacked. Lifting body pilots had concluded that the speed brake would assist energy management during the approach to landing. The M2-F3 pilots used the speed brake much like a throttle to make corrections to the approach speed and direct the lifting body to a precise landing point.

The rebuilt M2-F3 returned to the NASA Flight Research Center in July 1969.

Relocation of the Launch Switch

On August 21, 1969, Captain Jerry Gentry was inadvertently launched about 45 seconds early from the NB-52A in the X-24A. It can hardly be coincidence that the launch switch for the lifting bodies at the pilot's station in the NB-52B was relocated in September 1969. Twelve man-hours and seventy-six dollars were spent on the relocation of the switch. [80]

The M2-F3 hangs from the wing of the NB-52B in early 1971. It is suspended from he same pylon adapter as the X-24A. (NASA Photo E-22612)

The NB-52B carries the HL-10 to launch altitude in the company of a NASA Douglas F5D Skylancer chase plane. (Bob Dorr via Greg Spahr)

X-24A, M2-F3, and HL-10 lifting bodies sit on Rogers Dry Lake on December 18, 1969. (NASA Photo ECN-2353)

The HL-10 is suspended from its adapter on the X-15 pylon on the wing of the NB-52B. Vapor streams from the port on the side of the lifting body as the pilot primes the plumbing of the engine before launch on a powered flight. A Lockheed F-104 Starfighter flies chase off the right wing. (NASA photo)

Bill Dana glides the HL-10 to a landing on Rogers Dry Lake on November 17, 1969. (NASA Photo E21090)

John Manke stands in front of the HL-10 after its twenty-third flight on August 16, 1969. (NASA Photo EC69-2247)

Air Force Major Peter Hoag stands in front of the HL-10 after its twenty-sixth flight on September 30, 1969. (NASA Photo EC69-2247)

Fooling around. NASA pilots Bill Dana (left) and John A. Manke try to drag Air Force test pilot Peter Hoag away from the HL-10 lifting body while Air Force Major Jerauld R. Gentry is armed with a hammer in the cockpit on June 27, 1970. (NASA Photo ECN-2409)

Retirement of the NB-52A

The NB-52B's sister ship, NB-52A, 52-0003 was retired to the Military Aircraft Storage and Disposition Center (MASDC) at Davis-Monthan Air Force Base near Tucson, Arizona on October 19, 1969. This left the NB-52B as the sole mothership for the lifting body projects.

When the NB-52A was retired it had made just over 410 flights. It had launched the X-15-1 thirty times, the X-15-2 eleven times, and the X-15-3 thirty-one times. It had launched the M2-F2 four times, the HL-10 eleven times, and the X-24A twice. The NB-52A survives in the Pima Air Museum near Tucson, Arizona.

Simulating the Space Shuttle's Approach to Landing

In the first half of 1970, the NB-52B was used as an in flight simulator of the proposed Space Shuttle. Early studies of the Space Shuttle included retractable jet engines to provide power for flying the landing pattern at the end of each orbital flight. The jets were of no use to propel the shuttle into orbit and represented dead weight that reduced its available payload. The community of X-15 and lifting body pilots was of the opinion that an unpowered, steep angle of descent approach was a more appropriate means of landing the orbiters.

In much the same way that Lockheed F-104 Starfighters had been used to simulate the approach of the X-15, the NB-52B was used as an in flight simulator of the Space Shuttle. It was felt that the data derived from simulations of the low lift over drag (L/D) approaches by small aircraft like the Starfighter might not be applicable to reentry vehicles weighing 200,000 pounds or more. It was determined that the emergency descent flight mode of the B-52 Stratofortress would allow approaches to be flown at L/D ratios as low as three or four, similar to the L/D of the shuttle.

The NB-52B was used in conjunction with an F-111A to investigate the terminal area energy management of maneuvering reentry vehicles. Terminal area energy management is the process of ensuring that the vehicle arrives at the approach to the runway going the right speed. The F-111A was used to provide a comparison of the behavior of smaller aircraft performing steep, unpowered approaches.

The NB-52B made steep approaches to the runways on Rogers Dry Lake, simulating the approach to be used by the Shuttle orbiters. The pilots throttled six of the engines down to idle and set the remaining two engines to 75 percent thrust to provide power for the aircraft systems. The air brakes on the top of the wings were fully extended, to the "six" position. The NB-52B achieved lift-over-drag ratios as low as 3.2 in this configuration. The pilots pulled one and a half Gs during the flare for landing.

Six pilots flew fourteen straight-in approaches to the lakebed runways. The descents were started at 10,000 feet altitude. Eight overhead circling approaches were made, some full circles, others just three-quarters of a circle. The bank angle flown in the overhead approaches averaged forty-five degrees and reached as high as

The NB-52B simulates the diving approach of the Space Shuttle in early 1970. (NASA photo E21203)

The NB-52B carries the X-24A in 1970. (Bob Dorr via Greg Spahr)

The NB-52B is missing its nose art as it carries the X-24A. (AFFTC/HO)

The NB-52B hauls the X-24A to launch altitude on April 2, 1970. Major Jerry Bowline and Major Gordon Fornell launched John Manke on a flight that achieved mach 0.866 and a height of 58,700 feet. (AFFTC/HO)

X-24A drops away from the NB-52B in February 1971. (NASA Photo EC-2590)

sixty-five. Touchdown speeds varied between 185 miles per hour and 205 miles per hour.

Ground controlled approaches (GCA) were flown with the pilots of the NB-52B wearing hoods to restrict their view to the instruments inside the cockpit and taking instructions from the ground controller. Three pilots flew steep angle GCA approaches from 18,000 feet altitude, initiating the flare about 500 to 800 feet above the lakebed. Four such approaches were flown. The rate of descent on these approaches was so high that the vertical speed indicator was pegged at the maximum descent rate, denying rate of descent information to the pilots. The pilots judged that the task was not much more demanding than a standard ground controlled approach. These tests contributed to the conclusion that the jet engines should be removed from the Space Shuttle orbiters and substantially increased the payload of the Space Transportation System. [81]

Powered Lifting Body Flights

The NB-52B launched the HL-10 on the fastest lifting body flight on February 18, 1970. Captain Nicholas Fritz and Squadron Leader Allen Fisher launched Major Peter Hoag in the HL-10 on its thirty-fourth flight, which achieved a top speed of Mach-1.86 (1228 miles per hour). Victor Horton manned the launch panel operator's station.

The NB-52B launched the HL-10 on the highest altitude lifting body flight on February 27, 1970. Major Bowline and Captain George Larson launched Bill Dana in the HL-10 on its thirty-fifth flight, which attained an altitude of 90,303 feet. Jack Russell manned the launch panel operator's station.

The NB-52B launched the X-24A on its first XLR-11 rocket-powered flight on March 19, 1970. Lt. Col. Sturmthal and Major Bowline piloted the Stratofortress. Victor Horton manned the launch panel operator's station. Captain Jerauld Gentry flew the X-24A.

The Martin Marietta X-24A was displayed with the NB-52B at the Edwards Air Force Base Open house on May 19, 1970. The X-15 mission marks had been removed from the fuselage of the NB-52B, and a new series of profiles of the M2-F2, HL-10 and X-24A had been started.

The NB-52B first carried the M2-F3 on a captive-carry flight on May 22, 1970. Major Bowline and Squadron Leader Fisher piloted the mothership. Victor Horton manned the launch panel operator's station. Bill Dana occupied the cockpit of the M2-F3.

The NB-52B first launched the M2-F3 on a glide flight on

The NB-52B launches the X-24A on a rocket-powered flight in 1970. Note that the nose art has been removed. (NASA photo ECN2511)

Air Force Major Cecil Powell stands next to the X-24A following his first lifting body flight on February 4, 1971. (NASA Photo E-22715)

The X-24A was displayed next to the NB-52B at the Edwards Air Force Base open house on May 19, 1970. (Richard Lockett)

An Air Force Northrop T-38A Talon chases the NB-52B and M2-F3. Note that the left outboard engine nacelle has a black access panel from a B-52D. (Bob Dorr via Greg Spahr)

The NB-52B carries the M2-F3 on its original adapter on May 22, 1970. The nose art is missing, leaving a shiny circle. (U.S. Air Force Museum)

Bill Dana primes the engine of the M2-F3 shortly before launch on its twenty-fifth flight on August 25, 1971. Squadron Leader Allen Fisher and Major Dale Ranz piloted the Stratofortress. Victor Horton manned the launch panel. Dana flew the M2-F3 to a speed of Mach-1.095 (723 miles per hour) and an altitude of 67,300 feet. It was the first supersonic flight of the M2-F3. (NASA Photo ECN-2794)

The M2-F3 falls away from the NB-52B in 1972. Note that the buff-colored NB-52B chin radome has been replaced with a black radome taken from a B-52D. (NASA photo ECN3334)

Bill Dana in the M2-F3 drops away from the NB-52B on August 25, 1971. (NASA Photo EC71-2774

June 2, 1970. Squadron Leader Fisher and Captain Fritz flew the Stratofortress. Victor Horton manned the launch panel operator's station. Bill Dana piloted the M2-F3 on its first flight since Bruce Peterson's accident on May 10, 1967.

In June 1970, as the M2-F3 was being fueled prior to a ground test of its XLR-11 engine, the ground crew noticed water alcohol fuel flowing out of the liquid oxygen vent on the lifting body. It was immediately apparent that fuel and liquid oxygen had mixed in the liquid oxygen tank as a result of a failure of a valve in the rocket engine. Any ignition source or a sharp shock could detonate the liquid oxygen tank in the M2-F3. The ground crew carefully opened the vent valves on the liquid oxygen tank and

Bill Dana flares the M2-F3 just before landing of Rogers Dry Lake in December 1971. (NASA Photo EC-71-2930)

Bill Dana watches the NB-52B fly over after a flight in the summer of 1970. Note the vehicle parked across the runway centerline in the distance to emphasize that the runway is temporarily closed. (NASA Photo E-21628)

M2-F3 pilots Air Force Captain Jerauld Gentry, NASA pilot John Manke, NASA pilot William H. Dana, and Air Force Major Cecil Powell on March 5, 1971. (NASA Photo E-22901)

John Manke stands in front of the M2-F3 following its last flight on December 20, 1972. It was the twenty-seventh flight of the M2-F3. (NASA Photo ECN-3448)

The Northrop HL-10 was displayed next to the NB-52B at the May 17, 1972 Edwards AFB Open House, which marked the twenty-fifth anniversary of the Air Force. All mission marks had been removed from the NB-52B fuselage. (Richard Lockett)

retreated from the site. Nobody approached the aircraft for several days until it was certain that all of the liquid oxygen had boiled off. The Flight Operations Office cancelled all supersonic flights over the base to prevent the possibility that the impulse of a sonic boom might detonate the fuel/oxygen mixture. [82]

The liquid oxygen tank in the M2-F3 was purged of all traces of the water alcohol fuel, and the faulty valve in the engine was replaced.

Conclusion of the HL-10 Program

The HL-10 made the last of its thirty-seven flights on July 17, 1970. Major Fornell and Major Bowline launched Major Hoag from the NB-52B. Jack Russell manned the launch panel operator's station.

The NB-52B launched the HL-10 twenty-three times. The HL-10 survives on a pylon in front of the NASA Dryden Flight Research Center at Edwards Air Force Base.

The NB-52B launched the X-24A on its highest altitude flight on October 27, 1970. Lt. Col. Reidenauer and Squadron Leader Fisher launched John Manke on the nineteenth flight of the X-24A, which achieved an altitude of 71, 400 feet. Victor Horton manned the launch panel operator's station.

The NB-52B launched the X-24A on its fastest flight on March 29, 1971. Major Edward McDowell and Squadron Leader Fisher launched John Manke on the twenty-fifth flight of the X-24A, which achieved a speed of Mach 1.60 (1,036 miles per hour). Victor Horton manned the launch panel operator's station.

The NB-52B launched the X-24A on its last flight on June 4, 1971. Major McDowell and Major Ranz launched John Manke on a flight that reached a speed of Mach 0.817 and an altitude of 54,400 feet. Jack Russell manned the launch panel operator's station. The X-24A was launched from the NB-52B on twenty-seven of its twenty-eight flights.

M2-F3 Emergency Landing on Rosamond Dry Lake

Bill Dana made an emergency landing on Rosamond Dry Lake in the M2-F3 on September 24, 1971. Squadron Leader Fisher and Lt. Colonel Sturmthal piloted the NB-52B. Victor Horton manned the launch panel operator's station. Shortly after Dana ignited the XLR-11 rocket engine, it malfunctioned and caught fire. Dana shut down the engine, jettisoned the remaining propellants and glided to a landing on Rosamond Dry Lake.

The NB-52B launched the M2-F3 on its fastest flight on

December 13, 1972. Major George Luck and Captain Michael King piloted the Stratofortress. Victor Horton manned the launch panel operator's station. The landing gear of the NB-52B stuck in the down position after take-off. Luck and King cycled the landing gear a couple of times and then continued the mission with the landing gear extended. Later in the flight, about twenty minutes before the launch of the M2-F3, they succeeded in retracting the landing gear. They launched Bill Dana in the M2-F3 on its twenty-sixth flight, which achieved a top speed of Mach 1.613 (1064 miles per hour).

Conclusion of the M2-F3 Program

The NB-52B launched the M2-F3 on its highest altitude flight on December 20, 1972. Major George Luck and Major Frederick Feidler launched John Manke in the M2-F3 on its twenty-seventh flight, which reached a peak altitude of 71,500 feet. It was the final M2-F3 flight. Jack Russell manned the launch panel operator's station.

The M2-F3 is now displayed in the National Air and Space Museum.

Martin-Marietta X-24B

The Air Force Flight Dynamics Library proposed that the X-24A could serve as the basis for a lifting body with an entirely new shape. The airframe of the X-24A was sheathed in an aluminum glove that gave it a flat bottom, a long, pointed nose and stub wings. The flatiron shape was expected to provide greater controllability at hypersonic airspeed. In its new configuration, it was re-designated X-24B. The X-24B was the first lifting body to be equipped with nose-gear steering. The newly configured X-24B returned to Edwards Air Force Base in early 1973.

The cockpit of the X-24B sat farther aft relative to the center of gravity than had been the case with the previous lifting bodies. The cockpit was below the leading edge of the X-15 pylon while it was suspended from the wing. This meant that the pilot of the X-24B could not eject while the lifting body was mated to the wing of the NB-52B. Any emergency that threatened the NB-52B would require the lifting body to be launched before its pilot could eject.

Wind tunnel tests of the separation of the X-24B from the NB-52B were not performed. The accumulated launch experience with the X-15s and lifting bodies supported confidence in the method. The appropriate carry angle for the pylon adapter used by the X-24B was determined from the data acquired in tests of the

The X-24B hangs from the wing of the NB-52B on March 3, 1973, shortly after its arrival at the NASA Flight Research Center. (NASA Photo E25683)

previous lifting bodies. [83]

The NB-52B appeared at the Edwards Air Force Base Open House with the X-24B parked in front of its right wing on May 16, 1973.

The NB-52B carried the X-24B aloft on its first captive-carry flight on July 19, 1973. Lt. Col. Dale Ranz and Captain Thomas Higgins piloted the NB-52B. Victor Horton manned the launch panel operator's station. John Manke occupied the cockpit of the X-24B.

John Manke piloted the X-24B on its first glide flight on August 1, 1973. Lt. Col. Ranz and Captain Higgins flew the NB-52B. Victor Horton manned the launch panel operator's station.

The NB-52B launched the X-24B on its fastest flight on October 26, 1974. Major Luck and Captain Paul Skabo launched Mike Love in the X-24B on its sixteenth flight. The X-24B reached a maximum speed of Mach-1.76 (1,163 miles per hour) on that flight.

The NB-52B appeared at the Edwards Air Force Base Open House with the X-24B parked under the pylon on December 1, 1974.

The X-24B was first displayed to the public next to the NB-52B at the May 16, 1973 Edwards AFB Open House. The HL-10 was displayed next to the M2-F3 some distance down the flightline. The NB-52B mission marks for the X-15, M2-F2, HL-10, and X-24A had been reduced to a single silhouette of each with the number of flights launched. New profiles for the flights of the M2-F3 had been added. (Richard Lockett)

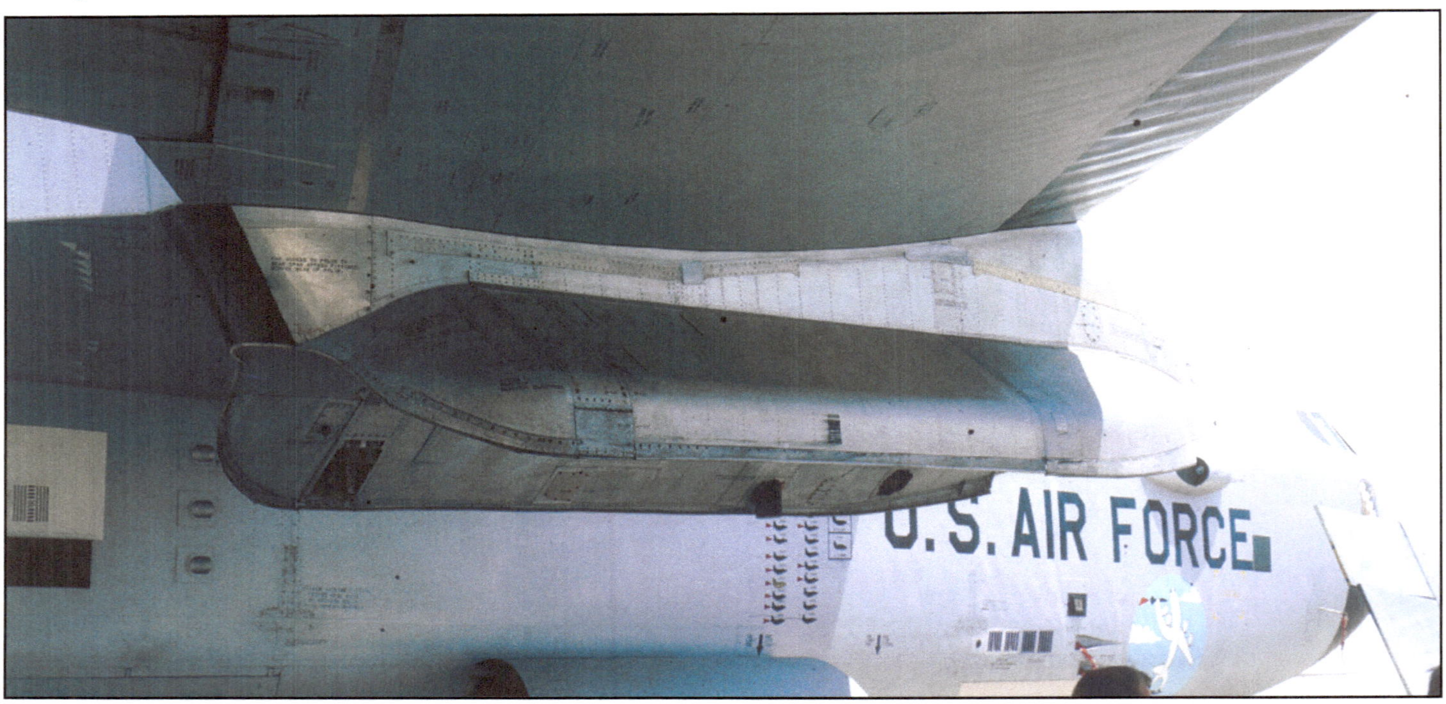

The X-15 pylon of the NB-52B on static display at the 1973 Edwards Air Force Base Open House. (Brian Lockett)

The NB-52B flying with the X-24B mounted on its wing in 1973. (NASA Photo ECN-3734)

The NB-52B carries the Martin-Marietta X-24B in 1973. (AFFTC/HO)

The NB-52B launches the Martin-Marietta X-24B in 1973. (Bob Dorr via Greg Spahr)

John Manke glides the X-24B to a landing on Rogers Dry Lake on its second flight on August 17, 1973. Lt. Col. Dale Ranz and Captain Thomas Higgins piloted the Stratofortress on that flight. Jack Russell manned the launch panel. (NASA Photo EC72-3742)

Mike Love stands next to the X-24B on Rogers Dry Lake in 1973. (NASA Photo E26487)

F-15 Remotely Piloted Research Vehicles

NB-52B operations entered a new phase whn Remotely Piloted Research Vehicles (RPRV) were introduced in the early 1970s. RPRVs pioneered a new experimental flight research technique that is now a common practice. RPRVs were promoted as a less expensive and safer alternative to manned flight-testing. The lightness and small size of RPRVs, compared with full-scale airplanes, allowed their construction to be quicker and less expensive. Pilots in control cockpits on the ground were not subject to the risks posed by flying aboard aircraft undergoing tests. [84]

The first RPRVs were three unpowered 3/8-scale models of the McDonnell-Douglas F-15 Eagle. The F-15 RPRVs were 7.15 meters long and had a wingspan of 4.89 meters. They weighed 1,135 kilograms. The RPRVs investigated the stall, departure, and spin characteristics of the F-15.

In addition to providing test data for the F-15 program, the scale models explored the validity of the RPRV concept itself. The F-15 RPRV aircraft investigated the correlation of data acquired from sub-scale flight research vehicles to that acquired from the full-scale aircraft. In effect, air is more viscous for a smaller airplane of the same design. This can have an effect on the behavior of the RPRV as compared to the full-scale aircraft. The F-15 RPRV

The NB-52B carries an F-15 RPRV in October 1973. (NASA photo ECN-3804)

program examined how well the behavior of a sub-scale research vehicle compared to that of the full-scale aircraft.

The pilot of the RPRV sat in a fixed-base simulator cockpit. He was provided with a video feed from a camera mounted in the cockpit of the drone. 50 channels of telemetry data were transmitted from the vehicle. The ground pilot displays included the attitude, angle of attack, angle of side slip, airspeed, altitude, g-load, control surface positions, and roll and yaw rates.

The operation of the RPRV's differed in significant ways from the full-scale aircraft. The arrangement of the instruments in the simulator cockpit was not the same as that in the full-scale airplane. The pilot was deprived of the motion cues that he would have experienced an actual flight. The field of view of the video camera was extremely narrow and did not provide peripheral vision to the pilot.

Pilots of the RPRV engaged in extensive simulation of the research missions between flights. The flight control laws of the simulator were tailored to allow the reproduction of the spin behavior of the RPRV. The research pilots practiced spin entry and recovery techniques for about ten hours for each five-minute research flight.

A typical F-15 RPRV mission began with launch from the NB-52B at an altitude of 50,000 feet and a speed of Mach-0.7. Tests were conducted during five minutes of gliding flight. About 10 stability and control maneuvers were performed on each flight. These maneuvers were repeated at several angles of attack to determine the RPRV's stability, control, and damping characteristics.

The F-15 RPRV's were equipped with stability dampers for all three axes. The function of the stability dampers was to prevent changes in the attitude of the drone that were not commanded by the pilot. For many of the flight test maneuvers, one of the stability damper channels would be disabled. This simplified the piloting task and allowed evaluation of the response of the aircraft to pilot control inputs to be limited to a single axis.

After completing the stability and control maneuvers, the

The 3/8-scale F-15 RPRV hangs from its adapter on the NB-52B pylon in April 1973. The pylon adapter is hoisted into place using the lift bar temporarily installed in the pylon. (NASA photo EC3644)

Recovery of the F-15 RPRV by a Mid-Air Retrieval System (MARS) helicopter was the standard method of operation for the first 16 flights of the program. The MARS helicopter snagged the drone as it descended under its parachute. (NASA Photo EC3925)

F-15 RPRV glides to a landing on Rogers Dry Lake on October 21, 1975. Major Thomas Higgins and Major Michael King piloted the Stratofortress. Victor Horton manned the launch panel. Major Francis Scobee controlled the F-15 RPRV from the ground. (NASA photo EC4882)

pilot would engage in one or two spin entry maneuvers. The RPRV descended approximately seven kilometers in the course of the spin testing.

It proved difficult to get to the F-15 RPRV to enter a spin. It was necessary to perform a precisely executed series of control inputs to spin the vehicle. Pilots practiced these maneuvers extensively in the simulator between flights.

The following example illustrates a typical spin research mission. To enter a spin, the flight controls were crossed. Full left rudder and full right aileron were applied. Once the spin was initiated, the elevator was centered. Thirty seconds after the maneuver was begun, the drone had spun eight times. The yaw rate reached 170 degrees per second and the angle of attack reached 75 degrees. Then the controls were neutralized and the yaw rate decreased to 100 degrees per second. The angle of attack decreased to 57 degrees. Full opposite control was applied to recover from the spin. The drone recovered after 6 seconds and two turns.

Recovery of the RPRV by the Mid-Air Retrieval System (MARS) helicopter was the standard method of operation for the first 16 flights of the program. When the RPRV had descended to an altitude of five kilometers, the pilot initiated the parachute recovery sequence. First, a small drogue parachute was released, followed by the main parachute. As the drone descended under the main parachute, a MARS helicopter snagged its shroud lines. The MARS helicopter delivered the drone directly to the NASA flightline.

Einar Enevoldson, John Manke, and Tom McMurtry were the F-15 RPRV program pilots.

The NB-52B first carried an F-15 RPRV on captive-carry flights on October 1 and October 10, 1973. Captain Higgins and Major Michael King piloted the Stratofortress on the first captive-carry flight. Captain Higgins and Captain Francis Scobee piloted the NB-52B when it first launched an F-15 RPRV on October 12.

On two occasions the MARS helicopter failed to retrieve the F-15 RPRV as it descended under its parachute. The first time, the drone landed using the main parachute, but sustained extensive damage. On the sixteenth flight, flown on January 16, 1975, the MARS helicopter failed to retrieve the RPRV. Tom McMurtry belly-landed the RPRV on Rogers Dry Lakebed. The damage suffered by the F-15 RPRV on those flights prompted a re-evaluation of the recovery method. Scheduling difficulties with the MARS helicopter provided additional motivation to seek an alternative recovery method.

After McMurtry had demonstrated the feasibility of landing the RPRV horizontally like an airplane, landing gear skids were installed on the drones. On subsequent flights the pilot flew the RPRV to a landing on the dry lakebed. On two of these flights, system problems forced the drone to descend to landing under the main parachute.

The limited field of view of the video feed made landing the RPRV an extremely demanding task. Researchers concluded that further development was required in the area of displays, additional landing cues for the pilot, and the quality of the television signal to relieve the workload of the pilot.[85]

Major Scobee guides the F-15 RPRV to a landing on its skids on Rogers Dry Lake on October 21, 1975. (NASA photo EC4890)

The NB-52B appeared at the December 1, 1974 Edwards AFB Open House in the company of the X-24B. The buff colored radome forward of the main gear has been replaced with a black one. (T. Waddington via Greg Spahr)

The NB-52B on July 6, 1975. (Brian Locket)

The LOX tank in the bomb bay of the NB-52B on July 6, 1975. (Brian Locket)

The X-24B adapter mounted on the X-15 pylon of the NB-52B on July 6, 1975. ((Brian Locket)

The X-24B lands on Rogers Dry Lake accompanied by NASA F-104N 811 in 1975. The rear half of the X-24B is painted white. (NASA Photo ECN-4351)

The NB-52B accelerates down Runway 04 with the X-24B before its twenty-seventh flight on August 5, 1975. John Manke performed the first runway landing of a lifting body on this flight. The failure of one of the rocket engine chambers limited John Manke in the X-24B to a maximum speed of mach 1.19 and an altitude of 57,000 feet as he flew from Cuddeback Dry Lake to Rogers Dry Lake. (NASA Photo E28846)

The NB-52B accelerates down Runway 04 with the X-24B. (Air Force Museum)

The flight control inputs of the pilots on these missions were recorded and fed into the flight simulator to determine how well the flight maneuvers were duplicated by the simulator. The flight control laws of the simulator were then adjusted to improve its ability to reproduce the flight of the drone.

Data collected on these missions was compared to wind-tunnel spin test data from 1/8 scale and 1/30 scale models. The spin characteristics of this design were essentially the same at all scales tested.[86]

Landing a Lifting Body on a Concrete Runway

The NB-52B launched the X-24B on its highest altitude flight on May 22, 1975. Major Luck and Captain Paul Skabo launched John Make in the X-24B on its twenty-third flight. Jack Russell manned the launch panel operator's station. The X-24B reached a maximum altitude of 74,130 feet on that flight.

The last abort of a manned launch from the NB-52B occurred on June 24, 1975. Major Higgins and Major Luck piloted the Stratofortress. Victor Horton manned the launch panel operator's

The NB-52B was accompanied by the X-24B at the November 16, 1975 Edwards AFB Open House. (Brian Lockett)

The X-24B sat next to the NB-52B at the November 16, 1975 Edwards AFB Open House. The beam in the front of the pylon is used to support a block-and-tackle rig for attaching the adapter. The X-24B made two more glide flights after this picture was taken. (Brian Lockett)

The NB-52B carries the X-24B in 1975. (U.S. Air Force Museum)

The NB-52B launches the X-24B on a glide flight in 1975. (Bob Dorr via Greg Spahr)

station. John Manke was scheduled to make the twenty-fifth flight of the X-24B. The take-off was delayed by twenty minutes due to a problem with the ailerons of the NB-52B. Once aloft it was determined that the wind was blowing across the planned landing runway, so Manke had to ride back down on the wing of the Stratofortress. All subsequent X-24B missions were launched successfully.

Lifting body flight research demonstrated that a vehicle with a low lift over drag ratio could be glided to a landing by using a steep angle of descent to keep airspeed high enough to maintain control until the pre-landing flare. This revealed that the space shuttle, under development at the time would not need to carry jet engines for the approach to landing. The deletion of the jets, their fuel, and the mechanical systems associated with them allowed a much larger fraction of the gross weight of the orbiter to be devoted to payload.

What the lifting bodies had not yet demonstrated was a landing in the confined space of a conventional runway. Because they all lacked a functional nose-wheel steering system, they required the large area available on Rogers Dry Lake to contain

The NB-52B launches the X-24B on a rocket powered flight in 1975. (NASA Photo EC75-4638)

The X-24B glides to a landing on Rogers Dry Lake in 1975. This view shows the facilities arrayed along the edge of the lake from North Base at lower right to the main runway at upper left. (NASA Photo EC75-4642)

the landing run-out.

The X-24B featured a functional nose-wheel steering system. This made it possible for the pilot to keep the X-24B lined up with the concrete runway during the landing rollout. The desired degree of precision was a touchdown within five hundred feet of a specified point on the runway. A five-foot wide white line painted across the main runway indicated the touchdown point to the pilot.

To prepare for landing the X-24B on the main runway, John Manke practiced steep approaches to the runways on Rogers Dry Lake in an F-104 Starfighter. Manke flew approximately 50 approaches to landing during the three weeks before the launch date.

Thomas C. McMurtry stands in front of the X-24B on Rogers Dry Lake in 1975. (NASA Photo ECN-4922)

On August 5, 1975, Fitzhugh Fulton and Capt. Thomas Higgins launched John Manke in the X-24B from the NB-52B over Cuddeback Lake, about thirty-five miles east of the runway at Edwards Air Force Base. As the X-24B descended from the launch altitude of 45,000 feet, Manke flipped the four switches that ignited the thrust chambers of the XLR-11 rocket engine. Only three of the chambers fired. With only three-quarters thrust, the planned flight profile to Mach-1.5 and 71,000 feet altitude was no longer possible. Manke switched to the alternate profile for a three-chamber flight that ascended to 58,000 feet and reached Mach-1.23.

Manke performed the planned data and performance maneuvers as the X-24B closed the distance to the runway. The ground controller, Lt. Colonel Mike Love, also an X-24B pilot, talked Manke to the low key point at 24,000 feet where a 180 degree turn to final was started. During the turn to final, Manke dropped the nose of the X-24B to increase its airspeed to 300 knots indicated airspeed. Winds blowing down the runway pushed the X-24B above the glide-slope, so Manke extended the upper flaps to act as a speed brake and keep the X-24B pointing at the preflare aim point.

Manke extended the landing gear when the X-24B was about a thousand feet above the runway and about a thousand feet past the threshold. The left tire touched down eight feet in front of the white marker line, and the right tire touched down five feet beyond it. The precision of the approach of the X-24B was more than adequate to prove the feasibility of landing a space shuttle within the confined area of a runway. [87]

The NB-52B launched the last rocket propelled lifting body

The NB-52B flies over the X-24B as it sits on Rogers Dry Lake after a flight in late 1975. (NASA Photo E-29230)

flight on September 23, 1975. Major Higgins launched Bill Dana in the X-24B on a flight that reached a speed of Mach-1.157 and an altitude of 56,800 feet. It was Dana's last rocket flight and also the last flight powered by one of the four-chambered Reaction Motors XLR-11 rocket engines that had powered the Bell X-1s in 1947. It was not just the same engine design; it was the same engine, still in use after twenty-eight years.

Conclusion of the X-24B Program

The NB-52B launched the last lifting body flight on November 26, 1975. Major Higgins and Fitzhugh Fulton launched Tom McMurtry in the X-24B on a glide flight. Six pilots had flown the X-24B on twelve glide flights and twenty-four powered flights.[88] The X-24B is displayed in the National Museum of the U. S. Air Force.

The NB-52B carried the heavyweight lifting bodies a total of 164 times and launched them 127 times. The NB-52A launched a further 17 lifting body flights.

Conclusion of F-15 RPRV Program

The NB-52B last launched an F-15 RPRV on December 17, 1975. Major Higgins and Captain Scobee piloted the Stratofortress and Ray Young manned the launch panel. Einar Enevoldson piloted the F-15 RPRV from the ground station.

The three F-15 RPRVs flew a total of twenty-seven times in their original configuration. Later, the F-15 RPRVs served as the basis for the Spin Research Vehicles.

The NB-52B would not conduct another launch for a year and a half.

Ryan Firebee II

The NASA Firebee II program was intended to demonstrate the feasibility of launching a modified Teledyne-Ryan Firebee II BQM-34 target drone from the NB-52B as the basis for a low-cost airfoil technology research vehicle. Three wing designs were envisioned for the program. The first phase of testing would utilize the Instrumented Standard Wing (ISW) that was based on the conventional Firebee II wing. Two supercritical wing designs were prepared for testing on the drone; Aeroelastic Research Wing 1 (ARW-1) and Aeroelastic Research Wing 2 (ARW-2).

Firebee II, 72-1564 was assigned to NASA's test program. Unlike the F-15 RPRV, the Firebee II was powered by a Continental YJ69-T-406 jet engine. It was capable of operating at a greater distance from the ground control station. Initial captive-carry tests of the NASA Firebee II were conducted with a U. S. Navy Lockheed DC-130A Hercules.

A DC-130A Hercules carried Firebee II, 72-1564 on a captive-carry flight on November 25, 1975. A second captive flight was scheduled for December 19. Problems with the command and control systems on the F-104 that was used for emergency control of the drone in flight caused that mission to be aborted.

The DC-130A Hercules carried Firebee II, 72-1564 on another captive-carry flight on May 17, 1976.

Transfer to NASA

The Air Force planned to retire the NB-52B to the Military Aircraft Storage and Disposition Center at Davis-Monthan Air Force Base once the lifting body and F-15 RPRV programs were concluded in 1975. NASA had a continuing need for the NB-52B to launch RPRVs and parachute test systems, so the NB-52B was loaned from the Air Force to NASA on April 26, 1976. NASA added a yellow tail band with black stripes and the word NASA to the vertical stabilizer of the Stratofortress. From then on, its radio call sign was NASA Zero Zero Eight.

The NB-52B was flown to Tinker Air Force Base, Kansas for IRAN on July 7, 1976. It returned to Edwards Air Force Base on September 22, 1976. In October, the NB-52B was flown on an airspeed calibration flight to facilitate air traffic control altitude reporting.

The NB-52B was little utilized during the next eight months. It went aloft on four training flights between October 1976 and June 1977. Its next mission was drop testing the Shuttle Reusable Booster Drop Test Vehicle.

Space Shuttle Solid Rocket Booster Drop Test Vehicle

The parachutes used to recover Space Shuttle Solid Fuel Boosters were tested in drops from the NB-52B. Marshall Space Flight Center awarded a contract to Martin-Marietta to build a 48,000-pound Space Shuttle Solid Rocket Booster Drop Test Vehicle (SRB/DTV) for drops from the X-15 pylon.

On February 24, 1977, the launch system circuitry at the pilot's and launch panel operator's stations was modified and re-wired in preparation for tests of the Space Shuttle Solid Rocket Booster Drop Test Vehicle (SRB/DTV)[89]. The emergency launch switch at the pilot's station was relocated on April 29[90]. Additional modifications to the SRB/DTV monitor and control panel were performed on May 14, 1977[91]. The pilot's launch panel circuit was modified on August 10, 1977[92].

The NB-52B made the first captive-carry flight with the SRB/DTV on June 10, 1977. The objectives of the flight were to check the Drop Test Vehicle systems and to measure the loads on the front hooks supporting the DTV. During the captive-carry flight, an instrumentation failure occurred after only one data point had been collected, so a second flight was flown the same day.

Six drops were made in the first series of tests of the SRB/DTV.

The first drop of the SRB/DTV was flown on June 15, 1977. Fitzhugh Fulton and Don Mallick piloted the NB-52B and Victor Horton manned the launch panel. During the descent of the DTV, the last pyrotechnic charge, used to separate the ballast section of the vehicle, failed to fire.

The SRB/DTV sits on its transporter next to the NB-52B on May 23, 1977. The NASA tail band reflects the transfer of the NB-52B to NASA on April 26, 1976. (NASA Photo E32492)

The Shuttle Reusable Booster Drop Test Vehicle hangs from the NB-52B pylon on June 17, 1977. (NASA photo EC7515)

The NB-52B carries the SRB/DTV on its first captive carry flight on June 10, 1977. (NASA Photo EC7429)

The NB-62B carries the SRB/DTV on a captive-carry flight on June 10, 1977. (NASA photo EC7427)

The SRB/DTV descends under a trio of Shuttle Reusable Booster parachutes on January 29, 1979. (NASA Photo EC79-10168)

Firebee II drone hangs from the NB-52B pylon in 1977. (NASA photo EC7401)

The NB-52B carrying Firebee II, 72-1564 in the summer of 1977. (NASA Photo ECN-8113)

NB-52B carries the Firebee II drone in 1977. Donald Mallick and Fitzhugh Fulton piloted the Stratofortress. Ray Young manned the launch panel. (NASA photo EC8114)

Launch of Firebee II

The NB-52B first carried NASA's Firebee II, 72-1564 on July 22, 1977. Don Mallick and Fitzhugh Fulton piloted the Stratofortress. Ray Young manned the launch panel operator's station.

The NB-52B launched the Firebee II at an altitude of 20,000 feet on July 28, 1977. Bill Dana controlled the flight of the drone from the ground. It climbed to an altitude of 40,000 feet. A MARS helicopter retrieved the Firebee II at the conclusion of the flight.

SRB/DTV Drop 2

The second DTV drop, flown on August 4, 1977 was a proof test with one main parachute.

Before the NB-52B carried the DTV again, it conducted the first series of launches of the F-15 Spin Research Vehicle and a pair of wake vortex measuring flights.

Wake Vortex Testing

The Air Force Weapons Lab conducted a study of B-52 wake properties with the NB-52B in 1977. To make the wingtip vortices of the NB-52B visible, a smoke system was installed to stream smoke from the its left wingtip. Plumbing and valves, a tank, and a service panel for the smoke system were installed in September and October 1977.[93] A smoke control panel and a control panel for a tail-mounted camera were fabricated at the same time.[94] An SCSG-1 smoke generator was installed in the left wingtip and a 16mm camera was installed in the tail gunner's compartment in September 1977.[95] An immersion heater was installed in the wingtip smoke generator in November 1977 to improve the flow of the oil used to generate the smoke.[96]

Two wake vortex tests were conducted at Idaho National Engineering Laboratory (INEL) near Idaho Falls in the fall of 1977. The NB-52B flew past a tower with wind speed and direction sensors mounted at several levels. The Stratofortress made eight passes by the tower on October 17.

Spin Research Vehicle

The F-15 RPRVs served as the basis for the Spin Research Vehicles (SRV). A variety of nose configurations and anti-spin parachutes were tested on the SRV. SRV launch panel circuitry for the NB-52B was fabricated in August 1977.[97] Einar Enevoldson, John Manke, and Mike Swann were the F-15 SRV project pilots.

The first attempt to launch an F-15 SRV in November 1977 was aborted by a hardware failure in the RPRV facility. The NB-52B first launched an F-15 SRV on November 29, 1977. The SRV flew a 40-turn spin as it descended from 44,000 feet to 23,000 feet in 90 seconds.

The NB-52B was displayed at the Edwards Air Force Base Open House on November 13, 1977, between the two wake vortex study flights. (Brian Lockett)

A Spin Research Vehicle was displayed on the pylon of the NB-52B on static display at the November 12, 1978 Edwards AFB Open House. (Brian Lockett)

A Spin Research Vehicle was displayed on the pylon of the NB-52B on static display at the November 12, 1978 Edwards AFB Open House. (Brian Lockett)

NB-52B carries Spin Research Vehicle in 1981. (EC81-15480)

Wake Vortex Flight 2

On December 6, 1977, the NB-52B made a refueling stop at Mountain Home Air Force Base and then flew twenty-two more passes by the Idaho National Engineering Laboratory air-data tower.

Spin Research Vehicle Flights 2 and 3

The NB-52B took off on two attempts to launch the second SRV mission on December 8, but both attempts were aborted. On December 9, the second F-15 SRV mission was launched at 52,000 feet, over a mile higher than the heavy rocket planes that the NB-52B had launched earlier. The SRV entered a spin to the right and flew a 73-turn spin.

A third SRV mission was flown on December 12, 1977 to test the addition of an anti-spin nose trip strip.

SRB/DTV Drops 3 Through 6

The third DTV test was conducted on December 14, 1977. The deployment of the drogue chute was delayed to evaluate whether it would survive the higher loads imposed when the DTV was going faster at lower altitude. The drogue chute failed.

The DTV failed to separate from the pylon of the NB-52B during an attempt to launch the fourth drop test on April 27, 1978. The fourth test was successfully launched on May 23. Problems with the J57 jet engines of the NB-52B delayed the fifth drop test until July 26. It duplicated the drogue chute overload conditions of the failed third drop test.

The sixth and last drop of the series was flown on September 12, 1978. It successfully demonstrated the deployment and structural integrity of the main chute cluster at the maximum expected operating load.

Fire Extinguishers

A pair of carbon dioxide fire extinguisher bottles was installed in the crew compartment of the NB-52B on June 1, 1978 [98].

The orange H-cradle supported the F-111 PTV in the bomb bay of the NB-52B. It is a relic of the days of free-fall hydrogen bombs. The NB-52B was the last Stratofortress with the capability of carrying the H-cradle. (NASA photo ECN18228)

Spin Research Vehicle Flights 4 Through 9

The NB-52B launched six F-15 SRV flights in September and October 1978. An air data boom was added to the nose of the SRV for these tests. The first two flights evaluated nose vortex flows. The next four flights investigated the effectiveness of an anti-spin helical trip strip, a 22-inch strake, and a nose mounted spin recovery parachute.

The next mission for the NB-52B was a drop of the F-111 Parachute Test Vehicle in February 1979.

F-111 Crew Module Parachute Test Vehicle

The Parachute Test Vehicle (PTV) was developed in the Sixties during the development of the original F-111 parachute system. The design of the PTV allowed it to be suspended in the bomb bay of any B-52 equipped for carrying large, free-fall thermonuclear weapons.

Over the operational history of the F-111, new equipment was frequently added to the cockpit. This increased the weight of the crew escape module. As a result, the rate of descent under the original seventy-foot diameter ring-sail parachute had increased, as had the risk of injury to a crew forced to eject from a disabled F-111. New parachute recovery systems were designed for the escape module of the F-111 to reduce the rate of descent to acceptable levels.

In the interval, the military B-52 fleet was modified to carry conventional bombs or missile launchers in the bomb bay. In the late Seventies, the NB-52B was the only remaining Stratofortress with a bomb bay configured to drop the PTV.

The installation of the equipment used to drop the PTV in the bomb bay of the NB-52B required the removal of the X-15 LOX top off tank. The tank and its associated cradle were removed on January 23, 1979 [99]. An H-frame supporting a standard MAU-12 bomb rack was installed in the bomb bay in February 1979 [100].

The PTV was ten feet long and four feet wide with two compartments for the test parachute installation and a safety

The Parachute Test Vehicle suspended in the bomb bay of the NB-52B. (NASA photo ECN18234)

The F-111 PTV drops from the bomb bay of the NB-52B. (NASA photo EC88-0032-03)

NB-52B on an F-111 PTV drop mission. (NASA Photo EC88-0008-06)

parachute system. It was pointed in front and had a drag parachute mounted on the back end to stabilize the vehicle until the test parachute deployed. It was painted bright red to make it easy to find on the desert floor.

The test parachute was attached to the top of the PTV during the initial deceleration. Once the PTV reached terminal descent, the attachment to the top of the vehicle was released. The load was taken up by fittings on the aft end of the vehicle. The pointed front end absorbed the force of impact with the ground to allow the reuse of the vehicle.

If the test parachute failed, the excessive rate of descent would automatically trigger its release, and the safety parachute would be deployed.

The parachute links were equipped with force sensors. The data from the sensors was stored onboard the vehicle in the Stored Data Acquisition System (SDAS) rather than being telemetered to the ground during the descent, which had been the case in earlier tests. The PTV was equipped with four 16mm movie cameras to record various aspects of the opening of the parachutes.

Eight drops were conducted in the first PTV test series. The drops were conducted over the Tonopah Test Range in Nevada. The NB-52B first dropped the PTV on February 17, 1979. The Stratofortress dropped the PTV three more times in February and March 1979.

Blue Streak

In 1979, the Instrumented Standard Wing was attached to Firebee II, 72-1564. It had a blue stripe painted along its span and was referred to as the "Blue Streak" wing. The NB-52B made a captive-carry flight of the Blue Streak on March 7, 1979. Fitzhugh Fulton and Donald Mallick piloted the Stratofortress and Ray Young manned the launch panel.

The NB-52B launched the Blue Streak only once, on March 9, 1979. Donald Mallick and Thomas LeBeau piloted the Stratofortress and Ray Young manned the launch panel. Tom McMurtry piloted the Blue Streak from the ground station.

F-111 PTV Drops 5 and 6

The NB-52B resumed F-111 PTV drop tests in May 1979. On the fifth PTV drop, the main parachute system malfunctioned. The PTV impacted the desert range at high velocity, and the box buried itself in the desert. NB-52B pilot Gordon Fullerton commented, "Surprisingly, it wasn't hurt much. It was built like a battleship, that test box, and we used it for later tests."

On the sixth drop, the test parachute deployment was inadvertently delayed, subjecting the parachutes to higher loading than intended.

The last two drops of the first F-111 PTV test series were accomplished as planned in August 1979. Before those drop tests, the NB-52B launched the first North American Rockwell HiMAT flight in July.

Blue Streak drone equipped with white Instrumented Standard Wing hangs from the NB-52B pylon in March 1979. (NASA photo ECN-10247)

The M2-F3 pylon adapter was modified to support the HiMAT drones for launch from the NB-52B. The main structural tubing of the adapter was converted to contain 230 gallons of jet fuel to replace the fuel burned by the HiMAT before launch. (NASA Photo EC11067)

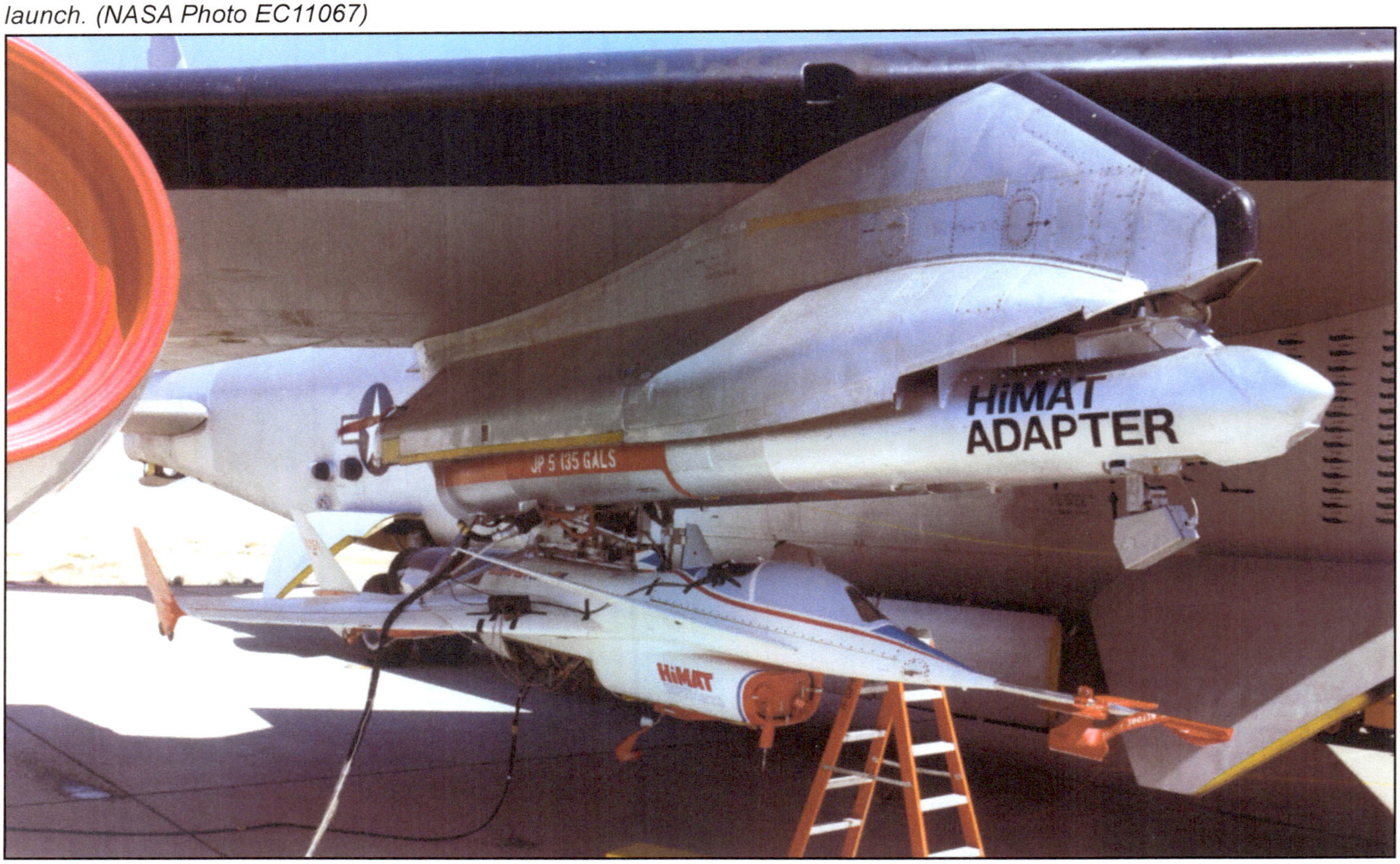

The HiMAT was mounted on the adapter in a 2-degree nose down attitude to ensure that it would separate cleanly from the NB-52B at launch. (NASA Photo EC11543)

Highly Maneuverable Aircraft Technology Drones

Initially RPRVs were used as an adjunct to the testing of a full-scale aircraft under development. Later RPRV's were developed to investigate the flight characteristics of new features that might be applied to future aircraft.[101] The success of the sub-scale F-15 Spin Test Vehicle, Firebee, and DAST operations from the NB-52B stimulated interest in the creation of research vehicles designed specifically for launch from the mothership. Department of Defense and NASA studies suggested that the maneuverability of fighters was a suitable subject of research for such a vehicle. The Highly Maneuverable Aircraft Technology (HiMAT) research vehicles were the result.

The HiMAT research vehicles were designed to investigate methods to improve the maneuverability of fighter airplanes. They were designed specifically for launch from the NB-52B. Novel aerodynamic features of the HiMAT drones included close-coupled canards and winglets. The drones were equipped with digital fly-by-wire and digital electronic engine controls in early tests of those technologies. Rockwell International designed a full-scale maneuverability demonstrator and then scaled it down to forty-four percent of its original size. Two HiMAT drones were built.

The close-coupled canards and winglets in conjunction with aeroelastic tailoring reduced the turning radius of the aircraft by half compared to contemporary fighter aircraft. The HiMAT design was capable of turning well inside current United States Air Force and Navy fighters.

The HiMAT drones were among the first flight vehicles manufactured from graphite epoxy composite materials. Aeroelastic tailoring involves controlling the alignment of the fibers in the various layers of the composite structures of the vehicle. The composite construction allowed the drone structures to be tailored for differing rigidity in different directions. The wings were strong and unbendable along their span, but they were flexible from leading to trailing edge. Flight control was accomplished by bending the entire wing surface using internal actuators. The wing had no hinge lines to interrupt the smoothly curved surface.

A natural result of the use of a sub-scale vehicle was a restriction on the space available inside the airframe for instrumentation. Space limitations demanded increased emphasis on the miniaturization of the electronics and instrumentation of the vehicle.

The empty weight of the HiMAT drones was 2,840 pounds. Fully fueled for launched they weighed 3,500 pounds. They were designed for a maximum speed of Mach-1.6. They were built to withstand twelve positive Gs and six negative Gs. A General Electric J85-21 turbojet engine powered each of the drones.

The M2-F3 pylon adapter was modified to support the HiMAT drones for launch from the NB-52B. The main structural tubing of the adapter was converted to contain 230 gallons of jet

The NB-52B taxis out for the first flight of HiMAT #1 on July 27, 1979. Fitzhugh Fulton and Thomas LeBeau piloted the Stratofortress. Ray Young manned the launch panel. Bill Dana controlled the flight of the HiMAT from the ground station. The tail of the 747 Shuttle Carrier Aircraft is just visible to the left of the drone. (NASA Photo EC11563)

fuel to replace the fuel burned by the HiMAT during the taxi, take-off, and climb to launch altitude. It could provide the drone with fifteen gallons of fuel per minute. The HiMAT was mounted on the adapter in a 2-degree nose down attitude to ensure that it would separate cleanly from the NB-52B at launch.

Two hooks on the adapter attached to pins on the centerline of the upper fuselage of the drones. Sway braces on the adapter prevented the drone from swinging from side to side. The pilot of the NB-52B controlled the release of the hooks. Compressed nitrogen gas drove the hook release mechanism.

Initial modifications to HiMAT panel wiring were made to the NB-52B in March 1978.[102] Additional wiring and a Volt-Amp monitor panel were installed in May.[103] Panel boxes, a T/M J-box, camera brackets, antennae, and a pneumatic launch system monitor were installed in January 1979.[104105] A relay was installed to shut down power switching during pyrotechnic payload loading.[106] The NB-52B power distribution was expanded for the HiMAT mission in April 1979.[107] At the same time, power kill circuitry was added to the electrical back-up launch system.[108] Circuitry associated with the fuel tank, camera, and power kill system was installed at the launch panel operator's station in May.[109] The fuel top off tank was equipped with a pressure sensor and a fuel pressure gauge was installed on the launch panel operator's instrument panel.[110] In November 1978 pylon connector wiring was rerouted to eliminate the pulling apart of hot wires from the connector in the event of an emergency launch of the HiMAT drone.[111]

The HiMAT engine was started on the ground before the NB-52B began to taxi to the runway. The NB-52B carried the HiMAT drone to an altitude of 45,000 feet for launch over Edwards Air Force Base.

The HiMAT drone was controlled in flight by a test pilot seated in a ground cockpit at the NASA Dryden Flight Research Center. Pilot control inputs were relayed to the HiMAT and processed by the onboard computer to generate the appropriate signals for the flight control actuators. The flight responses of the drone were downlinked to the ground cockpit displays. The test pilot was provided with a video feed from a forward-looking television camera. In the event that control of the drone from the ground was lost, a backup pilot in a NASA TF-104G Starfighter could take over control of the drone.

Test pilots found that commanding the HiMAT to perform a steady, high-G turn without changing altitude was more difficult than a similar maneuver in a manned airprlane. Motion and peripheral vision cues that were present in manned flight were not available to the test pilot in the ground cockpit.

It was determined that control over the flight test maneuvers could be improved by the introduction of a flight test maneuver autopilot. The autopilot allowed the ground pilot to put the drone into a particular flight attitude, which was then maintained by the autopilot to ensure stable flight conditions for recording data. The flight test maneuver autopilot has subsequently found application in manned flight test programs. It improves the quality of the data collected by maintaining flight conditions accurately and repeatably.

At the conclusion of each test flight, the drone was flown to a horizontal landing on skids on Rogers Dry Lake. Landing a HiMAT drone on the lakebed depending on the narrow field of view provided by the television feed proved to be a difficult task. Surprisingly, all HiMAT landings were well performed and no damage was inflicted on the drones during landing.

Bill Dana, Einar Enevoldson, and Steven Ishmael were the HiMAT program pilots. Bill Dana flew the first nine flights of the HiMAT drone. Ishmael joined the program in 1981 and Enevoldson joined in 1982.

The NB-52B made the first captive-carry flight of HiMAT #1 on July 11, 1979. Fitzhugh Fulton and Donald Mallick piloted the Stratofortress. Ray Young manned the launch panel operator's station. Bill Dana operated the HiMAT drone systems from the ground.

The NB-52B first launched HiMAT #1 on July 27, 1979. The crew was the same as for the captive carry flight, except Thomas LeBeau was the co-pilot of the NB-52B. Bill Dana flew the drone from the ground station. Control of the HiMAT was temporarily transferred to the backup control system in the airborne Lockheed TF-104G Starfighter due to a problem with the uplink from the ground station. The flight lasted 22 minutes.

Phase V Inspection

The NB-52B flew to Tinker Air Force Base, Oklahoma for its periodic Phase V inspection in August 1979. It returned to Edwards Air Force base in September.

F-16 Radar Target

The NB-52B served as a large radar target for General Dynamics F-16 Fighting Falcon targeting studies on at least one flight in 1979.

Drone for Aeroelastic Structure Testing

The Drone for Aeroelastic Structure Testing (DAST) was based on the airframe of the Ryan Firebee II drone. It was equipped with smooth surfaced Aeroelastic Research Wings with flexible skins and internal actuators instead of traditional hinged, fixed-contour flight control surfaces. The objective was to delay the onset of stall by eliminating discontinuities in the surface contour of the wing, and to provide complete control over wing camber to provide optimal lift at all speeds, altitudes, and maneuvering conditions.

The NB-52B was modified in preparation for the DAST program. Wiring for the drone was routed from the bomb bay to the launch panel operator's station and installed from the pylon to the bomb bay through the liquid oxygen plumbing on January 13, 1977[112]. A Firebee system console was installed at the launch panel

The NB-52B taxis out for the first flight of HiMAT #1 on July 27, 1979. Fitzhugh Fulton and Thomas LeBeau piloted the Stratofortress. Ray Young manned the launch panel. Bill Dana controlled the flight of the HiMAT from the ground station. The tail of the 747 Shuttle Carrier Aircraft is just visible to the left of the drone. (NASA Photo EC11563)

The NB-52B flies with DAST on a captive carry flight on September 14, 1979. (NASA Photo EC11693)

NB-52B carrying HiMAT #1 in 1979. Tufts of yarn are attached to the HiMAT to reveal the airflow around drone. (NASA photo EC11643)

The unpainted HiMAT #1 sits on Rogers Dry Lake. Note the vectored thrust engine nozzle. (NASA Photo)

The NB-52B sits on the flightline at the Dryden Flight Research Center in February 1980. (Bryan Wilburn via Bob Dorr)

operator's station on February 16, 1977[113].

The Firebee II drone was fitted with Aeroelastic Research Wing No. 1 (ARW-1) in 1979. The first captive-carry flight of DAST was flown on September 14, 1979. The NB-52B carried DAST around the same ground track planned for its free flight to check the radio signal strength between the DAST and the ground station. The NB-52B first launched DAST on October 2, 1979. A radio receiver failure prevented the collection of any of the flight test data points, so the flight was terminated early, and the MARS helicopter snagged the drone as it descended under its parachute.

HiMAT #1 Flights 2 and 3

The HiMAT program resumed flight operations in November 1979. The team aboard the NB-52B consisted of Fitzhugh Fulton, Don Mallick and Ray Young.

On November 14, the second flight of the drone was aborted shortly after its engine was started due to low pressure in the backup hydraulic system.

The failure of an accelerometer that was intended to detect the onset of flutter in the left winglet resulted in a mission abort two minutes before the next attempt to launch the HiMAT on December 20. The NB-52B launched the HiMAT drone on its second flight the next day. The flight lasted thirty-seven minutes. Four times during the flight the drone lost the uplink from the ground control station and control was transferred to the backup flight control system on the airborne TF-104 Starfighter.

The NB-52B launched the third HiMAT flight on January 15, 1980. Once again, control of the drone transferred to the backup control system in the Starfighter several times during the flight.

DAST Flights 2 and 3

Due to the loss of data on the first flight of DAST, the flight data collection system was modified for subsequent flights. The modifications were tested during a captive-carry flight on March 10, 1980. The NB-52B carried the DAST around the course that it would fly on its own to evaluate the signal strength of the telemetry signal at all points along the flight path.

The second DAST flight successfully collected the programmed flight data points at speeds up to Mach-0.9 on March 12, 1980. Before the next flight of the DAST, the NB-52B carried the HiMAT drone on a captive-carry mission on May 30.

The third flight of DAST was flown on June 12, 1980. Fitzhugh Fulton and Donald Mallick piloted the NB-52B. Ray Young manned the launch panel and Tom McMurtry flew the DAST from the ground. At a speed of Mach-0.825, the right wing of DAST fluttered and separated from the drone. It tumbled out of control and impacted in the desert. The MARS helicopter flew to the impact site and retrieved the severely damaged drone.

The occupant of the back seat in NASA Lockheed TF-104G Starfighter, N824NA took over control of the drone in the event of the loss of the control signal from the ground station. (NASA photo EC14095)

Planform view of DAST in flight on June 12, 1980 reveals the shape of the supercritical-airfoil Aeroelastic Research Wing number 1. Fitzhugh Fulton and Donald Mallick piloted the Stratofortress. Ray Young manned the launch panel. Tom McMurtry controlled DAST from the ground station. Shortly after this picture was taken, the right wing fluttered and separated from the drone. (NASA Photo EC80-14090)

HiMAT Flights 4 Through 8

The NB-52B launched the first HiMAT drone five times between June and December 1980. On June 24, the NB-52B carried the drone on a captive-carry flight to evaluate discrepancies that had been noted during the captive-carry mission on May 30. The HiMAT drone was launched on its fourth flight on June 25. Its jet engine flamed out one minute before landing. Bill Dana re-ignited the engine and landed successfully on Runway 15 on Rogers Dry Lake. The fifth flight of the HiMAT drone ended with a gear up landing on the lakebed on July 8.

The NB-52B made a functional check flight in August 1980 following an engine change. A pilot training and proficiency flight was conducted in September.

Three launch aborts preceded the next HiMAT flight. The sixth flight of the HiMAT drone was launched on October 28. Bill Dana stabilized the drone at Mach-0.9 at an altitude of 25,000 feet and then entered a sustained 7-g turn.

The first attempt to launch the seventh flight of the HiMAT drone was aborted four minutes before launch when it was discovered that the battery that powered the destruct system was dead from a shorted battery cell.

The HiMAT drone was flown successfully again on December 3 and December 18, 1980.

F-15 SRV Flights 10 through 21

The NB-52B spent the first half of 1981 launching a series of twelve flights of the F-15 SRV. The last previous SRV flight had been flown in October 1978. Einar Enevoldson flew the drone during first nine flights of the series.

The series of flights began inauspiciously when the anti-spin parachute mounted on the nose of the drone caught on the pitot tube during the tenth SRV flight on February 18, 1981. The drone failed to return to the dry lakebed and landed on the Edwards Air Force Base bombing range.

The air data boom was removed from the F-15 SRV and a longer nose was installed on the drone for its thirteenth flight on April 24. The SRV flew five flights in this configuration.

A strake was added to the long nose for the eighteenth flight of the F-15 SRV on June 1. The drone flew two missions in this configuration. Mike Swann flew the F-15 SRV for the first time on the nineteenth flight on June 5.

The long nose was replaced with a cambered nose for the twentieth SRV flight on June 15. A second flight in this configuration was flown on June 17.

Introduction of HiMAT Drone #2

The first captive-carry flight of the second HiMAT drone was scheduled for June 23, 1981. Fitzhugh Fulton and Don Mallick piloted the NB-52B. Ray Young manned the launch panel operator's station. Bill Dana operated the systems of the drone from the ground station. A hydraulic leak in the flight control system resulted in a postponement of the flight until June 25. The captive-carry flight revealed problems with the deployment of the landing gear, the operation of the backup control system, and the engine afterburner light sequence.

Conclusion of F-15 SRV Program

The NB-52B launched the F-15 SRV on its five last missions in July 1981. The drone was modified with a forebody vortex control nozzle installation. Jets of gas were expelled to delay the onset of a stall by preventing the formation of vortices on the nose of the drone during flight at high angles of attack. Two flights were flown in this configuration on July 2 and July 7.

Before the next flight, 0.0014 grit was glued to the surface of the lower right quadrant of the nose of the SRV to evaluate the effect of the texture change on airflow aft of the nose. Two flights were flown in this configuration on July 13 and July 14.

The NB-52B launched the last flight of the SRV on July 15, 1981. Fitzhugh Fulton and Brian O'Hara piloted the Stratofortress. Ray Young manned the launch panel and Mike Swan controlled the SRV from the ground. The nose of the drone was modified with a fifteen-degree cone ahead of the vortex control nozzles.

The Spin Research Vehicles flew twenty-six flights. Between the F-15 RPRV project and the SRV project, the NB-52B carried the three airframes a 73 times and launched them 53 times.

First Two Flights of HiMAT #2

On July 21, 1981 , six days after the last F-15 SRV flight, the NB-52B took the second HiMAT drone on another captive carry mission. All of the discrepancies that had been noted on its first captive-carry flight were resolved.

The NB-52B first launched the second HiMAT drone on July 24, 1981. Major Daniel Vanderhorst and Don Mallick piloted the NB-52B. Ray Young manned the launch panel operator's station. Bill Dana controlled the drone from the ground station. He collected airspeed calibration data at altitudes of 40,000 feet and 25,000 feet and speeds of Mach-0.7 and Mach-0.9.

Steven Ishmael was the second pilot to fly the HiMAT drone. He piloted the second HiMAT drone on its second flight on July 30, 1981. Seven minutes into the flight, the drone's jet engine started to malfunction, so the research mission was concluded and Ishmael landed the HiMAT on the lakebed.

Phase Inspection

The NB-52B flew to Kelly Air Force Base, Texas in August 1981 for scheduled phase inspection. While it was there, it was modified to bring it into compliance with outstanding technical orders for B-52s. After the inspection and modifications, the NB-52B returned to Edwards Air Force Base on September 4, 1981.

HiMAT Drones

Bill Dana and Steven Ishmael each flew the second HiMA

The NB-52B carries HiMAT #1 in 1980. (NASA photo ECN-14283)

HiMAT #1 in flight in December 1980. Fitzhugh Fulton and Don Mallick piloted the Stratofortress. Ray Young manned the launch panel. Bill Dana controlled the flight of the HiMAT from the ground station. (NASA Photo ECN-14273)

The NB-52B appeared with the HiMAT adapter on the X-15 pylon at the October 5, 1980 Edwards AFB Open House. (Brian Lockett)

drone once in the fall of 1981. A rear landing gear truck of the NB-52B failed to retract completely after take-off on September 18. Don Mallick and Major John Small piloted the Stratofortress. Ray Young manned the launch panel operator's station. Ishmael piloted the drone. The drone's left main landing gear partially extended fifteen minutes into the flight, so the mission was terminated early.

Bill Dana evaluated the sideforce-to-rudder feedback system of the first HiMAT drone on October 20.

The first HiMAT drone returned to the program with the first tests of Relaxed Static Stability (RSS) in December 1981. General aviation and airline airplanes have positive stablity and tend to return to straight and level flight. The HiMAT drones conducted early tests of a flight control system that allowed the airplane to be designed for neutral or even slightly negative stability for better maneuverability.

Steven Ishmael piloted the first HiMAT drone on its ninth flight on December 21, 1981, just over a year after its last previous flight.

Bill Dana demonstrated the HiMAT design goal of a sustained 8-g turn at a speed of Mach-0.9 and an altitude of 25,000 feet during the tenth flight of the first HiMAT drone on February 3, 1982.

Preparations for supersonic HiMAT flights delayed the next flight until May. The NB-52B began a second series of drops of the F-111 PTV.

F-111 PTV Second Series, Drops 1 Through 5

The NB-52B conducted a second test series of eleven PTV drops in 1982. The first five drops were conducted in March and May. The objective of the test series was to validate the ability of the parachute system to safely lower the F-111 crew module at a safe speed following an increase in its weight.

In preparation for the new drop tests, the NB-52B went aloft in February 1982 to check the operation of its bomb bay doors.

The first drop of the second test series was flown on March 3, 1982. Fitzhugh Fulton and Don Mallick piloted the NB-52B. Ray Young manned the launch panel operator's station. The emergency backup parachute deployed at the same time as the test parachute, so the test had to be repeated.

The NB-52B dropped the F-111 PTV twice more in March. Both drop tests failed due to a shroud line entanglement problem.

The conditions of the failed first drop were repeated during the fourth drop of the series in May. The fifth drop was configured to evaluate a fix for the entanglement problem revealed by the earlier tests. The deployment system was assembled with a

The NB-52B on static display at the Edwards Air Force Base Open House on September 20, 1981. (Brian Lockett)

The F-111 PTV damaged the bomb bay door latch release cable as it exited the bomb bay on a test drop in 1982. (NASA photo E39398)

simulated entanglement.

The next drop of the second F-111 PTV test series did not take place until October. In the meantime, the NB-52B launched a series of HiMAT drone flights.

HiMAT Drones go Supersonic

Nine HiMAT missions were conducted between May and September 1982. Fitzhugh Fulton and Don Mallick flew the NB-52B on most of the flights. Ray Young, Jerry Henry, and Harold O'Brien took turns manning the launch panel operator's station. Bill Dana and Steven Ishmael piloted the drones.

The first HiMAT drone flew the first two flights of the series. Steven Ishmael piloted the first supersonic HiMAT flight on May 11. Bill Dana successfully demonstrated the HiMAT's supersonic design performance goal on May 14.

Once the design performance goals for the HiMAT drones had been achieved, the program mission shifted to research. The second HiMAT drone made four flights for the purpose of evaluating pressures and loads imposed on the airframe of the drones during pushovers, pull-ups, and windup turns. Data collected included the amount of deflection of airframe components resulting from the aerodynamic loads.

The first HiMAT drone made two flights in August to investigate its stability at higher supersonic speeds with non-functional control surface dampers. Steven Ishmael flew the last flight of the first HiMAT drone on August 27. It was the fourteenth flight for that vehicle.

The NB-52B on static display at the Edwards Air Force Base Open House on October 23, 1982. (Brian Lockett)

Bill Dana flew the second HiMAT drone on September 14, but problems in the ground data recording system prevented the collection of data. A second attempt to collect the desired data on September 22 was aborted before take-off. Problems with the drone resulted in another launch abort on October 1.

Einar Enevoldson piloted a HiMAT drone for the first time on October 13, 1982. His flight continued the evaluation of loads and airframe deflections in high-g, supersonic flight.

Conclusion of F-111 PTV Second Test Series

During the month of October 1982, the NB-52B dropped the F-111 PTV four more times, completing the planned test series. The drop tests had revealed a potential problem with shroud line entanglement. Two additional tests were conducted in November to verify the effectiveness of modifications to the recovery system to alleviate the entanglement problem. The final drop of the second F-111 PTV test series was flown in November 1982.

DAST II

A second Aeroelastic Research Wing (ARW-1) was manufactured and installed on Firebee II, 72-1558. Additional avionics were installed in a bulbous fairing under the fuselage. The drone carried a 380-pound external fuel tank. The vehicle was dubbed DAST II, and prepared for flight in 1982.

Fitzhugh Fulton and Don Mallick piloted the Stratofortress on all DAST II missions. Tom McMurtry was the DAST II project pilot.

The NB-52B first carried DAST II on a captive-carry flight on October 29, 1982. Jerry Henry manned the launch panel operator's station. Tom McMurtry operated the systems of the drone from the ground station. The flight simulated the preparations for flight up to the moment of launch.

The NB-52B launched DAST II on November 3, 1982. Ray Young manned the launch panel on this and all subsequent DAST II missions. Tom McMurtry flew DAST II from the ground.

Conclusion of HiMAT Program

The NB-52B launched the second HiMAT drone on the last two flights of the program in January 1983. The penultimate HiMAT flight was first scheduled for November 23, but bad weather forced a postponement of the flight to the next day. The next attempt was aborted two minutes before launch due to a failure of the backup control system in the TF-104G Starfighter. A failure of the Aeronautical Test Range computer prevented the next attempt on December 22.

The second HiMAT drone was launched for the eleventh time on January 7, 1983. Einar Enevoldson piloted the drone despite the failure of the backup control system. Four days later a switch problem in the TF-104G Starfighter postponed the last HiMAT flight until the next day.

Einar Enevoldson made the last flight of the HiMAT program on January 12, 1983. Fitzhugh Fulton and Donald Mallick piloted the Stratofortress. Harold O'Brien manned the launch panel operator's station. It was the twelfth flight of the second HiMAT drone. Although another DAST II flight was scheduled soon after, this was to be the last launch of a remotely piloted vehicle from the NB-52B.

While the aerodynamic data recorded on HiMAT flights was of good quality, they did not provide an ability to assess the performance of the vehicle in air combat maneuvering. The remote piloting setup did not provide the pilot with any sense of the feel of the response of the aircraft to his inputs.

Initially, it was anticipated that the remotely piloted vehicles would be relatively expendable. It turned out that the cost and visibility of the program caused the loss of the flight vehicles to become less acceptable. Over the course of the F-15 RPRV, DAST, and HiMAT programs it was found that they crashed once for every 12 to 15 flights that they flew. The relatively high crash rate of these programs impacted the cost savings that had been anticipated.

The effort expended on ground operations, pre-flight, and flight was found to be equivalent to that expended on manned flight research programs. Dryden Flight Research Center people assigned 30 to 40 people to the F-15 RPRV program. This was about the same as the number assigned to the lifting body programs. Total support for the RPRV program increased to 80 people on flight days. This was equivalent to level of effort required for the lifting body programs.[114]

It was noted that although the remotely piloted research vehicles produce data of good quality, those programs often failed to excite the pilot community within Government and industry. Without a man in the cockpit, much of the allure of flight-testing is lost on most observers. To the public and the funding decision-makers in Washington, the operation begins to look a lot like flying model planes. While this does not impact the quality of the data, it can impact the availability of funding to conduct flight research.[115]

Conclusion of NB-52B support of DAST II

The second flight of DAST II was scheduled for January 20, but the jet engine of the drone overheated during the start procedure. The NB-52B took off with DAST II again on February 10, but the jet engine RPM indicator in the ground cockpit failed, and the mission was postponed again. A third launch attempt later in February was prevented by a failure of a rudder position sensor on the drone.

No further attempts were made to launch DAST II from the NB-52B. The NB-52B carried the Firebee II, Blue Streak, DAST, and DAST II a total of sixteen times and launched them six times. Two flights ended in crashes.

All subsequent DAST II launches were conducted from a U. S. Navy DC-130A Hercules. The Hercules carried DAST II on two captive-carry flights in May 1983. DAST II was launched from the DC-130A on June 1. The parachute recovery system deployed immediately after launch. The main parachute separated, and the drone crashed in a farm field near Harper Dry Lake. This mission is referred to as the "Alfalfa Impact Test. Although it was launched from the DC-130A, there is a mission mark on the NB-52B commemorating this flight.

The crash of DAST II ended the Firebee-derived drone program. The second Aeroelastic Research Wing (ARW-2) was never flown.

The NB-52B carries DAST II on a captive-carry flight on October 29, 1982. Fitzhugh Fulton and Donald Mallick piloted the Stratofortress. Jerry Henry manned the launch panel. Tom McMurtry controlled DAST from the ground station. (NASA Photo EC-20773)

The NB-52B carries DAST II on a captive-carry flight on October 29, 1982. (NASA Photo EC-20777)

NB-52B carries DAST II before its first free flight on November 3, 1982. Fitzhugh Fulton and Donald Mallick piloted the Stratofortress. Ray Young manned the launch panel. Tom McMurtry controlled DAST II from the ground station. (NASA photo E20830)

Close-up of the underside of DAST II on the wing of the NB-52B in 1982. (NASA Photo E20785)

Space Shuttle Solid Rocket Booster Drop Test Vehicle Ii

The introduction of a larger parachute for the recovery of Space Shuttle reusable solid-fueled boosters prompted a second series of drop tests of the SRB/DTV in 1983. Three large stabilizing fins were added to the Drop Test Vehicle for these drops. Eight drops were planned for the test series.

The first drop of the second test series was initially scheduled for February 23, 1983. The flight was cancelled due to adverse weather. As the NB-52B taxied back to the flightline, the rear hooks supporting the SRB/DTV failed. The resulting damage put the NB-52B pylon out of commission until September.

An analysis of the loads on the pylon hooks revealed the need for modifications. Over the summer of 1983 the hooks were modified and then tested to validate their load carrying ability.

The NB-52B dropped the first test of the enlarged parachute over the Navy's China Lake Test Range on September 16, 1983. The first three drops were conducted at an altitude of 20,000 and airspeed of 200 knots. The fourth drop of the test series validated a new drogue chute design on March 30, 1984.

The last four tests were dropped from 40,000 feet and continued the validation of the new drogue chute.

The fifth drop was scheduled for July 19. The drop was postponed for one day when weather conditions deteriorated in the drop zone after the NB-52B took off.

The sixth drop was successfully conducted on October 30,

The NB-52B launches DAST II on its first free flight on November 3, 1982. (NASA Photo EC20816)

U. S. Navy Lockheed DC-130A 570497 carries DAST II before launch on May 3, 1983. (NASA Photo EC24124)

The rear attachment point of the SRB/DTV failed as the NB-52B was taxiing on February 23, 1983. (NASA Photo EC22789)

The SRB/DTV rests on the ground after the failure of the NB-52B pylon hooks on February 23, 1983. (NASA photo EC22788)

The NB-52B hauls the SRB/DTV to the El Centro Range in the Imperial Valley. (NASA Photo via Greg Spahr)

The NB-52B hauls the SRB/DTV to the El Centro Range in the Imperial Valley on March 28, 1984. (NASA Photo EC28734)

The SRB/DTV falls away from the NB-52B on July 20, 1984. (NASA photo EC30263)

The NB-52B releases the SRB/DTV over the El Centro Range in the summer of 1983. (NASA Photo E39908)

but an anomaly occurred during the descent of the seventh drop on January 17, 1985. The SRB/DTV impacted the China Lake Test Range at very high speed. There is a special mission mark on the fuselage of the NB-52B showing the unit stuck in the ground and flanked by Joshua trees to commemorate the event.

The eighth drop concluded the second SRB/DTV test series on March 20, 1985.

Trip to Kelly Air Force Base

The NB-52B flew to Kelly Air Force Base, Texas in October 1985. An Air Force team replaced its main landing gear with units from a later model of Stratofortress. While it was at Kelly Air Force Base, it was placed on static display for an airshow.

The NB-52B returned to Edwards Air Force Base later in October. It was placed in flyable storage. It did not fly again until March of 1987.

The NB-52B resumed flying with a functional check flight in March 1987. NASA Ames-Dryden Flight Research Facility pilots who were scheduled to participate in the next series of F-111 PTV drop tests went aloft in the Stratofortress on five training flights in March. Space Shuttle astronaut Gordon Fullerton made his first flight in the NB-52B with Don Mallick on March 10, 1987.

The NB-52B at the Kelly air Force Base, Texas airshow in October 1985. (Cliff Bossie via Greg Spahr)

The NB-52B was in flyable storage when it appeared on static display at the Edwards Air Force Base Open House on November 9, 1986. (Brian Lockett)

F-111 PTV Drop Test Series 3

Sandia Laboratories developed an alternative cluster of three 52.5-foot diameter combination ringslot-solid parachutes for the F-111 crew escape module. The F-111 PTV was used to test these parachutes in a third series of drops conducted from July 1987 to November 1988.[116]

The parachute system was required to operate over a wide range of airspeeds. A pilot parachute large enough for a slow speed ejection would pull too hard for an ejection at high speed. To keep deceleration loads at acceptable levels, the solution was to use a seven-foot diameter pilot parachute with panels designed to blow out at high speed.

Early tests of the seven-foot diameter pilot parachute revealed that the blow out panels were not failing consistently, resulting in high loads on the main parachute deployment links. Later tests used a seven-foot diameter, low porosity, ripstop pilot parachute.

The NB-52B flew from Edwards Air Force Base to the Tonopah Test Range in Nevada to drop the PTV. Initial tests proved the deployment system of the test parachutes. Later tests were conducted at progressively higher airspeeds to increase the dynamic pressure, known as Q in engineering shorthand, to which the parachutes were subjected.

The NB-52B went aloft on a functional check flight on July 1, 1987 to prepare for the first drop of the PTV. The big red box was dropped from the Stratofortress at an altitude of 12,000 feet and airspeed of 170 knots on July 7.

The PTV test series was interrupted by the departure of the NB-52B to Kelly Air Force Base, Texas for Periodic Depot Maintenance in July. The Stratofortress returned to Edwards Air Force Base on September 12.

The second test duplicated the drop conditions of the first test on September 22, 1987. The safety parachute deployed at the same time as the test parachutes, rendering the test invalid.

The drop altitude was raised to 14,000 feet starting with the third test on October 23. The airspeed was raised to 200 knots for the fifth test on January 7, 1988. The airspeed was raised progressively to 225 knots, 250 knots, and 275 knots over the next four flights. During the first drop at 275 knots on May 4, 1988 there were "some anomalies" during the deployment sequence.

Smaller pilot and main parachutes were introduced for the tenth drop test on August 17. The pilot parachute was reduced to five feet in diameter and the main parachutes were forty-nine feet in diameter. The airspeed for the drop was reduced to 225 knots for the three tests of the smaller parachutes.

An entanglement caused the premature release of the six-

The F-111 PTV drops from the bomb bay of the NB-52B. (NASA photo EC88-0057-02)

The F-111 PTV drops from the bomb bay of the NB-52B on February 19, 1988. (NASA photo EC88-0032-03)

foot diameter guide surface-vent parachute, which is attached to the main parachutes, during the eleventh test on September 13, 1988. As a result, one of the main parachutes developed a false vent and failed catastrophically. The other two main parachutes failed in quick succession. The resulting high rate of descent triggered the release of the test parachutes and the deployment of the safety parachute system.

The twelfth and final test of the third F-111 PTV test series was performed on November 9, 1988, the day after the 1988 Edwards Air Force Base Open House. The final configuration tested used a five-foot diameter guide surface parachute inside a ten-foot diameter circular flat parachute. The larger parachute was rigged to break away if the drag force was too high, leaving the smaller parachute to deploy the main parachutes.[117]

Golden Bird

In 1988, the Naval Air Systems Command was conducting the Glint and Scintillation Program. The NB-52B was assigned to support the Naval Weapons Center "Golden Bird" test phase of the program in November 1988. A range problem resulted in the mission being aborted.

Call for Retirement of the NB-52B

In 1988 a NASA budget committee recommended the retirement of the NB-52B. The committee pointed out that committing the airplane to any future projects beyond 1990 would be "a significant gamble from a logistics point of view." Fortunately for the Stratofortress, its unique ability to carry and launch very large payloads made it indispensable as a launch platform for the upcoming Orbital Sciences Pegasus booster.

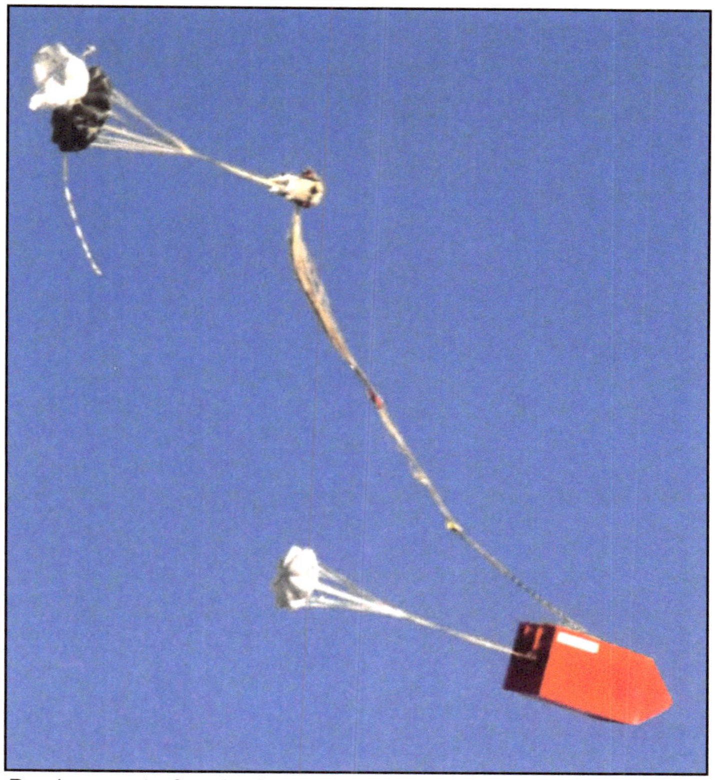

Deployment of a single parachute recovery system for the F-111 crew escape module on May 16, 1989. (NASA Photo EC89-0126-03)

Deployment of a single parachute recovery system for the F-111 crew escape module. (NASA Photo ECN89-308-10)

Launch crew for F-111 PTV test drops poses in front of the NB-52B in 1988. Space Shuttle astronaut Gordon Fullerton is at left. Ed Schneider is at right. (NASA Photo ECN88-030-05)

The NB-52B was displayed with a HiMAT drone attached to the pylon at the November 8, 1987 Edwards AFB Open House. (Brian Lockett)

The NB-52B on the flightline at NASA Dryden in May 1989. (Craig Kaston via Greg Spahr)

Orbital Sciences Pegasus

Under contract to the Defense Advanced Research Projects Agency (DARPA), Orbital Sciences Corporation designed the Pegasus booster to be air-launched from the NB-52B. The three-stage Pegasus booster weighed over 41,000 pounds.

Initial preparations for Pegasus launches from the NB-52B were begun in 1989. Stratofortress flights in support of the Pegasus launch mission were interspersed between drops of the F-111 PTV. A Pegasus launch rehearsal was flown in June 1989 to evaluate the timing of the launch sequence.

In August 1989, the NB-52B flew a survey of the test range in which Pegasus boosters were to be launched. The Stratofortress flew along the same ground track to be used for the upcoming launches to evaluate the test range systems and obtain accelerometer data.

F-111 PTV Drop Test Series 4, Drops 1 Through 9

A fourth series of eleven F-111 PTV drop tests began in May 1989. The first drop was conducted at and altitude of 14,000 feet and airspeed of 225 knots. The airspeed was raised to 275 knots for the second drop in June. The three main parachutes remained reefed and the PTV impacted the test range at high speed.

A month later, the third drop repeated the conditions of the failed second test. The three main parachutes inflated asymmetrically and failed.

The NB-52B dropped the PTV three times in August 1989.

An inert Orbital Sciences Pegasus booster hangs from the wing of the NB-52B in late 1989. (NASA Photo EC89371)

A new adapter was created to mount the Pegasus booster on the X-15 pylon. (NASA Photo EC89376)

The NB-52B on static display at the Edwards Air Force Base Open House on October 29, 1989. (Brian Lockett)

For the fourth test of the series, the drop altitude was raised to 18,600 feet and the airspeed to 359 knots. The deployment of the test parachute deviated from the programmed sequence. The fifth test of the series reproduced the launch conditions of the fourth test and the same deployment sequence problem cropped up again.

Later in August, the sixth PTV test of the series was dropped at an altitude of 14,000 feet and airspeed of 250 knots and displayed no problems during the deployment sequence.

The airspeed of the NB-52B was increased to 350 knots for the seventh test of the series in September. The three main parachutes inflated asymmetrically and failed. The next drop repeated the same conditions and experienced no problems.

The ninth drop of the fourth test series was flown in October 1989. The altitude of the drop was increased to 18,600 feet and the airspeed to 325 knots. The three-parachute cluster deployed successfully in the proper sequence.

Pegasus Taxi Tests and Captive Flights

The NB-52B engaged in taxi tests with an inert Pegasus booster in November 1989. Low-speed taxi tests reached a top speed of 47 knots on November 4. High-speed taxi tests reached a top speed of 110 knots on November 9.

Later on the same day as the high-speed taxi test, the NB-52B carried the inert Pegasus on a captive-carry flight, primarily to make sure that the Pegasus booster did not experience any flutter while attached to the wing of the NB-52B at the speeds and altitudes that would be encountered on an actual launch mission.

A second Pegasus booster captive-carry flight in December revealed problems with launch procedures and telemetry reception. A third captive-carry flight was flown on January 30, 1990 to rectify the issues.

F-111 PTV Drop Test Series 4, Drops 10 and 11

The fourth F-111 PTV test series concluded with two drops of a single parachute recovery system in December 1989 and February 1990. The airspeed for the tenth drop was raised to 335 knots. The test parachute separated from the PTV. The safety parachute also separated from the PTV and it impacted the ground at "a significantly higher velocity than planned".

The eleventh and last drop of the fourth PTV test series reproduced the same launch conditions as the previous mission. The single parachute deployed successfully on the final test.

The NB-52B carries an inert Orbital Sciences Pegasus on its first captive-carry flight on November 9, 1989. . (NASA photo EC89-0309-3)

First Orbital Sciences Pegasus Booster Launch

The NB-52B flew a crew proficiency flight with the Pegasus adapter mounted on the pylon in March 1990. The flight crew tested the function of the adapter hook safing pin and release systems.

The first Pegasus booster launch was scheduled for April 4, 1990, but the launch was postponed for a day because of clouds over the Pacific Ocean launch zone.

The NB-52B became the first airplane to launch a satellite into orbit on April 5, 1990. Gordon Fullerton and Jim Smolka piloted the NB-52B.[118] The Pegasus was launched at an altitude of 43,000 feet and airspeed of Mach-0.83. They launched the Pegasus booster over the Pacific Ocean, about 60 miles southwest of Monterey, California. It was the first time a payload had been put into earth orbit by a commercial rocket.

Space Shuttle Drag Chute Tests

The NB-52B tested the Space Shuttle Drag Chute in a series of eight high-speed taxi tests in 1990. NASA's Johnson Space Center contracted with Rockwell Space Systems to construct a mechanism to test the drag chute intended for use to decelerate landing Space Shuttle orbiters. A boxy structure was installed under the empennage of the Stratofortress to hold the drag chute mechanism.

The first four drag chute tests were conducted on Rogers Dry Lake. The tests moved to the concrete main runway for the last four tests.

The drag chute was deployed at an air speed of 150 miles per hour during the first test on July 20, 1990. Reefing lines restricted the size of the drag chute to 28% of its fully deployed size for several seconds after initial deployment. The braking force generated by the parachute was 19,000 pounds while reefed and 49,000 pounds when fully deployed.

The deployment airspeed was raised to 166 knots for the second drag chute test on August 2. The reefing ratio remained at 28%. The braking force generated by the parachute was 21,000 pounds while reefed and 61,000 pounds when fully deployed.

The deployment airspeed was raised to 176 knots for the third drag chute test on August 17. The reefing ratio was increased to 40%. The braking force generated by the parachute was 31,000 pounds while reefed and 65,000 pounds when fully deployed.

Shuttle Drag Chute tests moved to the main runway for the fifth test on September 12. Gordon Fullerton and Ed Schneider were at the controls of the NB-52B. The deployment airspeed was 163 knots. The braking force generated by the parachute was 30,900 pounds while reefed and 65,400 pounds when fully deployed.

The highest speed at which the NB-52B deployed the chute was 192 knots on the sixth test, conducted on September 21. The drag chute remained reefed to prevent it from separating from the Stratofortress. The braking force generated by the parachute was 44,300 pounds.

The drag chute was deployed without reefing at an airspeed of 153 knots during the seventh test on October 4, 1990. The braking force generated by the parachute was 56,600 pounds.

The last drag chute test simulated a failure of one of the two reefing lines. The braking force generated by the parachute was 29,900 pounds while reefed and 64,300 pounds when fully deployed. The drag chute system shear pin broke under the load and the parachute separated from the NB-52B.

The Space Shuttle Endeavor made the first use of the drag chute on May 16, 1992.

Orbital Sciences Inertial Navigation System Checkout

Orbital Sciences Corporation technicians checked out the operation of a reworked inertial navigation units on a flight of the NB-52B in December 1990.

F-111 PTV Drop Test Series 5

A new 95-foot diameter ring sail parachute system was developed for the F-111 crew module in the early 1990s. A series of four drops of the F-111 PTV were flown to test the new parachute

The NB-52B became the first airplane to launch a satellite into orbit when ti launched the first Pegasus booster on April 5, 1990. (NASA photo EC90-0111-2)

The NB-52B deploys the Shuttle Drag Chute for the first time on July 20, 1990. (NASA Photo EC90-210-1)

The NB-52B deploys the Shuttle Drag Chute for the first time on July 20, 1990. (NASA Photo EC90-225-30)

NB-52B tests the Space Shuttle drag chute on Rogers Dry Lake in August 1990. (NASA photo EC90-257-32)

The NB-52B deploys the Space Shuttle Drag Chute on Runway 22 on September 12, 1990. (NASA Photo EC90-262-28)

The NB-52B tests the Space Shuttle drag chute on the concrete runway on September 27, 1990, (NASA photo EC90-262-27)

in February and March 1991.

The NB-52B dropped the F-111 PTV at an altitude of 9,000 feet and airspeed of 160 knots during the first test of the series in February. The altitude was raised to 18,900 feet and the airspeed increased to 211 knots for the second test in March. The airspeed was increased to 336 knots for the third test. The fourth test repeated the launch conditions of the first test.

Preparations for second Pegasus Booster Launch

The NB-52B was flown on a series of functional check flights in May 1991. The flight crew checked out the operation of a Global Positioning Satellite receiver had been installed in the Pegasus adapter. They conducted a full rehearsal of the upcoming second Pegasus launch.

F-111 PTV Drop Test Series 6

A sixth series of ten F-111 PTV drops was conducted from June 1991 to November 1991. Low-altitude drops were conducted at 9,000 feet, and high-altitude drops were conducted at 18,600 feet.

These sixth series of PTV drop tests was combined with tests of a ground speed sensor that was to be used on NASA's Convair 990 Landing Systems Research Aircraft (LSRA). The LSRA had a Space Shuttle main landing gear truck mounted between its own main landing gear trucks. The LSRA made low passes across Rogers Dry Lake, touching down on the Space Shuttle landing gear truck to measure the loads imposed on it. A critical aspect of the tests was an accurate measurement of the ground speed of the airplane when the gear touched down. The ground speed sensor was tested on most of the flights made for the PTV drop tests.

The first drop test of the sixth series was conducted in June 1991. Following the PTV drop on the first five missions, the NB-52B tested the LSRA ground speed sensor. The second PTV test was a high-altitude drop at an airspeed of 211 knots. The third test was a low-altitude drop at an airspeed of 160 knots. For the fourth

The Shuttle Drag Chute was housed in the green box underneath the tail of the NB-52B, seen here on static display at the Edwards Air Force Base Open House on October 6, 1990. (Brian Lockett)

The NB-52B was displayed with a Pegasus satellite launcher at the October 6, 1990 Edwards AFB Open House. (Brian Lockett)

test was a high-altitude drop at an airspeed of 270 knots. The fifth test repeated the conditions of the third test.

The sixth PTV test drop of the series was a high-altitude drop with the airspeed increased to 336 knots. The parachute canopy failed. The airspeed for the rest of the test drops in the series was 336 knots. The seventh drop test was the last low-altitude drop of the series. The last three drops of the series were all high-altitude drops. The parachute shroud lines failed during the eighth test of the series.

Following the ninth test drop in October, the NB-52B made a flight devoted to a test of the LSRA ground speed sensor. Dust impaired the sensor readings. It was the last test of the ground speed sensor aboard the Stratofortress before it was installed in the LSRA.

The tenth and last drop of the sixth F-111 PTV test series was flown in November 1991. The launch conditions repeated those of the previous two drops.

Second Orbital Sciences Pegasus Booster Launch

The second Pegasus booster was launched during the sixth series of F-111 PTV drop tests. A full rehearsal of the Pegasus launch sequence was flown in early July 1991

The NB-52B launched the second Pegasus at and altitude of 43,000 feet and airspeed of Mach-0.83 on July 17, 1991, using the same drop point off the California coast as the first launch. The second Pegasus booster successfully placed seven Microsats into orbit.

The NB-52B on static display at the Edwards Air Force Base Open House on October 6, 1991. (Brian Lockett)

F-111 PTV Drop Test Series 7

A seventh series of five F-111 PTV drops began in March 1992. The drop conditions for the first test were the same as the last drops of the sixth series. The main canopy of the parachute failed. After the drop, one of the right-hand main landing gear trucks of the NB-52B failed to extend. The crew had to resort to the standby gear extension system to get it lowered. The F-111 PTV was dropped four more times in 1982, concluding the PTV test program.

The F-111 PTV program was the longest running program supported by the NB-52B, spanning a period of over thirteen years. The Stratofortress dropped the PTV from its bomb bay 61 times.

During the rest of 1992, the NB-52B went aloft on a series of crew proficiency and training flights. On one of these flights in September, the right-hand, front main landing gear truck would not extend. The crew had to use the standby extension system to lower the landing gear again.

The NB-52B on static display at the Edwards Air Force Base Open House on October 18, 1992. (Brian Lockett)

The NB-52B rolls down Runway 04 with a Pegasus launcher in 1994. (NASA Photo EC94891)

The NB-52B returns to Edwards AFB with the Pegasus adapter mated to the pylon following a satellite launch. A NASA F/A-18B is flying chase. (NASA Photo EC94-42695-1)

Orbital Sciences Pegasus Booster Launches 3 and 4

The third Pegasus booster launch carried Brazilsat, the first commercial satellite to be launched by Pegasus. In order to place the satellite in a low inclination orbit (an orbit that is near the equator) rather than a polar orbit, the launch was conducted over the Atlantic Ocean off of Florida.

The NB-52B took off to ferry the third Pegasus booster to the Kennedy Space Center in January 1993. Before the Stratofortress got as far as Barstow, the rudder of the Pegasus booster swung hard over. Although the rudder displacement was unnoticeable to the crew of the Stratofortress, concern that vibration on the long flight to Florida might damage the booster resulted in a return to Edwards Air Force Base. The NB-52B deployed to Kennedy Space Center with the Pegasus booster successfully in early February.

The NB-52B launched the third Pegasus booster approximately 90 miles east of the Kennedy Space Center on Feb. 9, 1993. Gordon Fullerton and Jim Smolka piloted the Stratofortress. The NB-52B operations engineer directed the launch operation from the control room at NASA's Wallops Flight Facility. According to Fullerton, as the operations engineer intoned, "Three, two, one, Launch!" over the radio, other personnel in the control room were hollering, "Abort, abort, abort!" on a separate communications loop. Fullerton later related, "The good news is the rocket did fine. The call really didn't need to be made. The desired orbit was attained and it all came out well, in spite of the confusion in the control room. That incident is classic now, on these CRM [Crew Resource Management] courses that have been presented by several training firms on how not to do control room ops."

The Brazilsat launch was the only Pegasus booster mission that the NB-52B launched out of the Kennedy Space Center.

The NB-52B launched the fourth, fifth, and sixth Pegasus missions off the California coast. The fourth Pegasus booster launched the ALEXIS satellite on April 25, 1993. Despite a successful launch, it proved impossible to establish communication with the satellite.

Supersonic Supercruise

In 1993 the NB-52B began a series of tests of fuel additives. A pair of J85 jet engines was mounted on a pallet that fit in the bomb bay of the Stratofortress in place of the center section of the bomb bay doors. The pallet incorporated tanks of specialized fuel additives. It is reported that the additives were intended to reduce emissions of oxides of nitrogen from the proposed High Speed Civil Transport. Emissions of oxides of nitrogen in the stratosphere would have a detrimental effect on the ozone layer.

Initial tests of the engine pallet were conducted in April 1993. For over two years the NB-52B flew as a ten-engine airplane. Mission marks on the side of the NB-52B record thirty-four flights of the ten-engine jet.

Ten-engine jet. The NB-52B tests fuel additives with a pair of J85 jet engines mounted on a pallet in its bomb bay on June 30, 1995. (Brian Lockett)

The NB-52B flying with a pair of J85 jet engines mounted on a pallet in the bomb bay. The X-15 pylon has been removed and a close-out structure has been installed in the slot in the trailing edge of the wing. (Brian Lockett)

A NASA F/A-18B Hornet chase plane accompanies the NB-52B on June30, 1995. (Brian Lockett)

Orbital Sciences' Lockheed L-1011, N140SC Stargazer, assumed the task of launching Pegasus missions. (Brian Lockett)

Dryden research pilots pose in front of NASA McDonnell-Douglas F/A-18B Hornet, N850NA in August 1995. From left to right: Thomas C. McMurtry, Edward T. Schneider, James W. Smolka, Dana D. Purifoy, Roger E. Smith, and C. Gordon Fullerton. All of these pilots flew the NB-52B in the 1990s. (NASA Photo EC95-43228-1)

Orbital Sciences Pegasus Booster Launches 5 and 6

The fifth Pegasus booster placed the Advanced Photovoltaic Electronics Experiment in January 1994.

The NB-52B launched its last satellite on the sixth Pegasus booster on August 3, 1994. The Stratofortress carried Pegasus boosters thirteen times and launched a total of seventeen satellites.

Pegasus launches have since been conducted from Orbital Sciences' Lockheed L-1011, N140SC *Stargazer*.

Removal of the X-15 Pylon

The X-15 pylon affected the airflow around the J85 engines use for the fuel additive tests. In order to eliminate the asymmetric flow field around the NB-52B, the 1,170-pound X-15 pylon was removed on September 15, 1994. On October 7, 1994, a close-out structure was installed to fill the slot cut in the trailing edge of the wing to accommodate the tail of the X-15.

Fortieth Anniversary

The NB-52B made two passes over the NASA Dryden Flight Research Center to commemmorate the fortieth anniversary of its first flight on June 30, 1995. A Dryden F/A-18 chase plane accompanied the Stratofortress as it flew over the facility.

Periodic Depot Maintenance

The NB-52B flew to Tinker Air Force Base in 1996 for Periodic Depot Maintenance. It returned to Edwards Air Force Base in November 1996.

X-38 Space Station Crew Return Vehicle Parachute Test-bed

The NB-52B conducted a series of seven drops of NASA's X-38 Space Station Crew Return Vehicle parachute test-bed. The X-38 was an 80%-scale technology demonstrator for the proposed crew return vehicle that was intended to serve as an autonomous "lifeboat" for the International Space Station. NASA planned to

The control panel of the NB-52B in 1996. The red backup release handle is visible at lower left. (NASA Photo ECN96-43814-01)

Initial mating tests of the X-38 V-131 were conducted in July 1997. (Brian Lockett)

X-38 V-131 is suspended from block and tackle as it is hoisted onto the new pylon on the wing of the NB-52B on July 28, 1997. (Brian Lockett)

The NB-52B appeared with X-38 V-131 mounted on its new pylon at the 1997 Open House. The Open House ran for two days, October 18 and 19. (Brian Lockett)

use the crew return vehicle to return the Space Station crew to Earth if an orbiter was unavailable or if the Space Station became disabled.

A new pylon was manufactured to carry the X-38 under the wing of the NB-52B. The new pylon weighed 3,000 pounds.

The original X-38 shape was derived directly from the Martin-Marietta X-24A. Its flying characteristics had already been thoroughly explored. Unlike the X-24A, which landed at over two hundred miles per hour, the X-38 landed gently under a large, wing-shaped parafoil.

The X-38 conducted tests of the last phases of the landing of the proposed Crew Return Vehicle. After launch from the NB-52B, it demonstrated the behavior of the lifting body shape in gliding flight at subsonic velocity, then it deployed a large parafoil for the final descent, landing very slowly on its skids on Rogers Dry Lake.

Two X-38 vehicles were constructed. V-131 was initially intended to make just short glide flights, so it lacked movable

The NB-52B carries X-38 V-131 in the company of NASA F/A-18B Hornet N846NA in 1997. (NASA Photo)

The NB-52B carries X-38 V-131 in the company of NASA F/A-18B Hornet N846NA in 1997. (NASA Photo)

flight controls. V-132 was equipped with functional flight control surfaces, which allowed it to perform extended glides before deploying the 5,000-square foot parafoil.

Merri Sanchez, an X-38 launch panel operator, was possibly the first woman to fly aboard the NB-52B on June 5, 1997

The NB-52B first carried X-38 V-131 on a captive-carry flight on August 2, 1997.

In addition to a pair of NASA F/A-18B Hornet chase planes, X-38 missions utilized the crimson Wolfe Air Cessna 337, N337WA as a low speed chase plane with a stabilized television mount. A Bell UH-1 Huey helicopter flew near the X-38 as it descended under its parafoil, transmitting live television of the test.

The first parachute drop test of X-38 V-131 was conducted on March 12, 1998. The parafoil shroud lines twisted as it deployed, and the lifting body oscillated from side to side for several seconds. The unanticipated behavior of the parafoil as it deployed prompted a ten-month hiatus in X-38 flight tests while the parafoil was tested at Yuma Marine Corps Air Station.

X-38 V-131 was dropped from the NB-52B for the second time on February 6, 1999. V-131 was subsequently modified by Scaled Composites to more closely represent the configuration of the orbital V-201 Crew Return Vehicle. After modification, its designation was changed to V-131R. It tested a larger 7,500-square foot parafoil intended for the full-scale vehicle.

The NB-52B first launched X-38 V-132 on March 5, 1999. V-132 made its second flight on July 9, 1999 and its third flight on March 30, 2000.

Ed Schneider, Frank Batteas, and David Dennis manned the NB-52B as it flew its one-thousandth flight on September 16, 1999.

The NB-52B first launched X-38 V-131R with the larger parafoil on the sixth X-38 mission on November 2, 2000. Shortly after the NB-52B took-off, 747 Shuttle Carrier Aircraft, NASA905 departed for the Kennedy Space Center with Shuttle Orbiter Discovery. The X-38 was dropped from the Stratofortress at an altitude of 36,500 feet. It made an unplanned 360-degree roll after launch. After twenty-four seconds of gliding flight, it deployed an 80-foot diameter drogue chute. It deployed the 7,000 square foot parafoil at an altitude of 19,000 feet.

The NB-52B took up position behind a Boeing KC-135 Stratotanker. A photographer in the boom operator's station aboard the Stratotanker shot a portrait of the Stratofortress that appeared on the cover of the Smithsonian Institution's *Air & Space* magazine.

The seventh X-38 mission was scheduled for June 29, 2001. Just seconds before drop, a failure of the flight termination system resulted in a launch abort. The NB-52B circled to make another attempt to drip X-38 V-131R, but the McDonnell-Douglas F/A-18B chase planes did not have enough fuel to accommodate the

X-38 V-131 rides on the wing of the NB-52B during a captive carry flight on November 19, 1997. (NASA Photo EC97-44319-8)

X-38 V-132 falls away from the NB-52B on July 9, 1999 Note the chromate green close-out structure completely fills the trailing edge slot in the wing of the NB-52B. (NASA photo EC99-45080-25)

X-38 V-132 sat on its trailer on static display next to the NB-52B at the 1999 Open House, which was held on October 9 and 10. (Brian Lockett)

The NB-52B on static display at the Edwards Air Force Base Open House on October 21, 2000. (Brian Lockett)

NB-52B takes off with X-38 V131R on November 2, 2000. (Brian Lockett)

NB-52B takes off with X-38 V131R on November 2, 2000. (Brian Lockett)

The NB-52B carries X-38, V-131R to the drop point on November 2, 2000. (NASA Photo EC00-0317-154)

The NB-52B was photographed from a Boeing KC-135 Stratotanker on March 30, 2001. The close-out structure in the trailing edge of the Stratofortress has been modified to make room for the vertical stabilizer of the X-43A Hyper-X stack. (NASA Photo EC01-0084-5)

delay. The mission was postponed until July 2.

A power failure shut down the systems of the control room shortly before the scheduled take-off of the NB-52B on July 2. The take-off was postponed for two hours. The Lockheed-Martin X-35 Joint Strike Fighter prototype kicked up dust from Rogers Dry Lake as it ran up its engine in the hours before the X-38 mission. The seventh X-38 mission was conducted with "no deltas" shortly before noon.

The Stratofortress launched the last X-38 flight on December 13, 2001.

The Space Station Crew Return Vehicle and the X-38 flight test program were cancelled in late 2001. The NB-52B carried the X-38s a total of twenty-one times and launched them eight times. X-38 V-132 was placed on static display at the Johnson Space Center.

Technicians prepare X-38 V-131R for a mission on June 29, 2001. (Brian Lockett)

The crew of the NB-52B starts up eight J57 jet engines, ten air driven hydraulic packs, and four alternators before taxiing out for a mission on June 29, 2001. (Brian Lockett)

X-38 V-131R hangs from the wing of the NB-52B on June 29, 2001. *(Brian Lockett)*

The NB-52B takes-off from Runway 22 with X-38 V-131R on June 29, 2001. *(Brian Lockett)*

The NB-52B takes-off from Runway 22 with X-38 V-131R on June 29, 2001. The mission was aborted just seconds before launch. *(Brian Lockett)*

The NB-52B drops X-38 V-131R on July 10, 2001. A smoke generator has been installed in the left inboard engine nacelle. (NASA Photo EC01-0204-2)

The NB-52B drops X-38 V-131R on its eighth and last mission on December 13, 2001. (NASA Photo EC01-0339-33)

X-38 V-131R descends under its 7,500-square-foot parachute to its final landing on Rogers Dry Lake on December 13, 2001. (NASA Photo EC01-0339-146)

Frank Batteas has flown the NB-52B on many flights. He is seen here in a NASA McDonnell Douglas F/A-18 Hornet. (NASA photo EC99-44835-1)

X-43A Hyper-X

The NB-52B launched three X-43A Hyper-X stacks to investigate Supersonic Combustion Ramjet technology. The X-15 pylon was re-installed on the wing of the Stratofortress for the Hyper-X program. The twelve foot long X-43A was mounted on the first stage of a Pegasus booster and dropped from the NB-52B. After launch the Pegasus booster pitched up into a climb as it does during a typical satellite launch, but then it pitched over to level flight at about 100,000 feet altitude and accelerated to the desired launch speed. The X-43A separated from the booster and successfully demonstrated the operation of a supersonic combustion ramjet at a speed approaching Mach-10.

NB-52B with the first X-43A during the first captive carry flight on April 28. (NASA Photo EC01-0126-06)

NB-52B with the first X-43A during the first captive carry flight on April 28. (NASA Photo EC01-0126-07)

The original X-15 pylon with the Pegasus adapter was re-installed for Hyper-X launches. (Brian Lockett)

The X-43A and its Pegasus booster were mated to the NB-52B for the first time on March 13, 2001. The NB-52B first carried an X–43A Hyper-X stack on a captive-carry flight on April 28, 2001.

The first X-43A Hyper-X mission was launched on June 2, 2001. The Pegasus booster had more than enough propellant to reach the desired airspeed and altitude. In order to consume the full load of propellant, the stack was dropped at an altitude of 20,000 feet, rather than the typical Pegasus launch altitude of 40,000 feet. As the Hyper-X stack accelerated in the denser atmosphere, the increase in dynamic pressure relative to the normal Pegasus launch altitude overloaded the rudder actuator. The vehicle yawed out of control and had to be destroyed by range safety control.

NB-52B takes off with the first X-43A Hyper-X stack on June 2, 2001. (Brian Lockett)

NB-52B takes off with the first X-43A Hyper-X stack on June 2, 2001. (Brian Lockett)

NB-52B takes off with the first X-43A Hyper-X stack on June 2, 2001. (Brian Lockett)

The NB-52B launches the first X-43A Hyper-X stack on June 2, 2001. (NASA Photo EC01-0182-20)

The first X-43A Hyper-X stack tumbles out of control on June 2, 2001. (NASA Photo EC01-0182-11)

Hyper-X Hiatus

The failure of the first X-43A Hyper-X stack resulted in a lengthy hiatus in the program. Nearly three years passed before the next X-43A flight. In March 2002, the NB-52B and NASA's newly acquired white and blue NB-52H, 61-0025 flew to the Boeing plant in Wichita, Kansas for a celebration of the fiftieth anniversary of the first flight of the Boeing YB-52.

The NB-52B was displayed, along with the second X-43A Hyper-X vehicle, inside the large hangar at the east end of the Edwards Air Force Base flightline during the 2002 Open House. The Stratofortress was displayed on the flightline at the 2003 Open House.

The NB-52B was displayed at a celebration of the 50th anniversary of the first B-52 flight at the Boeing plant at Wichita, Kansas on April 14, 2002. NASA's new NB-52H, 61-0025 was also displayed at Wichita. (NASA Photo EC02-0080-38)

The NB-52B was displayed indoors at the Edwards Air Force Base Open House on October 26, 2002. (Brian Lockett)

The second X-43A Hyper-X vehicle was displayed at the Edwards Air Force Base Open House on October 26, 2002. (Brian Lockett)

The NB-52B on static display at the Edwards Air Force Base Open House on October 24, 2003. (Brian Lockett)

Hyper-X Flight II

After a lengthy investigation into the loss of the first X-43A Hyper-X stack, it was determined that excessive aerodynamic loads at the launch altitude of 20,000 feet had overloaded the rudder actuator on the modified Pegasus booster. The rudder actuator stopped working, allowing the vehicle to yaw out of control. In short order the stabilizing fins were ripped away and the vehicle tumbled out of control.

The solution to the problem was to launch the X-43A Hyper-X stack at 40,000 feet, reducing the solid-fuel propellant load to limit the top speed of the booster to Mach-7. The Pegasus booster was modified by the removal of 3,300 pounds of excess propellant.

The second X-43A Hyper-X was taken aloft on one captive carry flight and then launched on March 27, 2004. The X-43A separated from the booster at a speed of Mach-7. The inlet door to the engine was opened and fuel was injected into the combustion chamber. The engine ignited and produced thrust that exceeded the drag of the Hyper-X for several seconds. The second X-43A demonstrated that an air-breathing engine could propel a vehicle at Mach-7. It was the first flight of an integrated supersonic combustion ramjet powered vehicle and shattered all previous records for the speed of a vehicle propelled by an air-breathing engine. The X-43A accelerated in a climb at a speed more than twice as fast as the Lockheed SR-71A Blackbird.

The NB-52B on static display at the Edwards Air Force Base Open House on October 24, 2003. (Brian Lockett)

NASA F/A-18B Hornet, N846NA accompanies the NB-52B as it carries the second X-43A Hyper-X stack on March 27, 2004. (NASA Photo EC04-0092-18)

The second X-43A Hyper-X stack accelerated to a speed of Mach-7 on March 27, 2004. (NASA Photo EC04-0092-32)

The NB-52B flies of Dryden Flight Research Center following the successful Hyper-X mission on March 27, 2004. (NASA Photo EC04-0092-43)

The third X-43A Hyper-X vehicle on March 27, 2004. (Brian Lockett)

NASA's Boeing NB-52B Stratofortress Mothership takes off on its last flight with the third X-43A Hyper-X stack on November 16, 2004. (Brian Lockett)

NASA's Boeing NB-52B Stratofortress Mothership climbs out steeply from Runway 04 with the third X-43A Hyper-X stack on November 16, 2004. (Brian Lockett)

Final Mission

The final mission of the NB-52B was the launch of the third X-43A Hyper-X on November 16, 2004. Former Space Shuttle pilot C. Gordon Fullerton piloted the Stratofortress mothership. The co-pilot was Frank Batteas. Orbital Sciences employee Brian Minnick manned the launch panel and NASA Dryden's Brad Neal monitored the X-43A systems aboard the NB-52B.

The Pegasus booster climbed to 110,000 feet and accelerated to Mach-9.8 in about 90 seconds and then released the X-43A. Once it was established that the X-43A was in stable flight, the engine cowl doors opened. Hydrogen fuel was injected into the combustion chamber and ignited with Silane. Combustion was maintained in the airflow going faster than the speed of sound through the engine. The thrust of the engine was approximately equal to the drag of the vehicle, simulating cruise flight, for ten seconds. The X-43A flew twenty miles in those ten seconds.

The Stratofortress carried the three X-43A Hyper-X vehicles a total of six times and launched them three times.

The final launch: the NB-52B launches the third X-43A Hyper-X stack on November 16, 2004. (NASA Photo EC04-0325-32)

The NB-52B makes a pass over the NASA Dryden Flight Research Center following the last X-43A Hyper-X mission on November 16, 2004. ((Brian Lockett)

Into the sunset: a pair of NASA F/A-18B Hornet chase planes fly in formation with the NB-52B as it makes a final pass over the Dryden Flight Research Center. (Brian Lockett)

The NB-52B sits outside the main hangar at the NASA Dryden Flight Research Center on the day of its retirement ceremony, December 17, 2004. (Brian Lockett)

Attendees at the retirement ceremony for the NB-52B on December 17, 2004. (Brian Lockett)

The NB-52B had flown 1,051 times and accumulated 2,440 flight hours when it landed for the last time at the end of the third X-43A Hyper-X mission. Over the course of nearly forty-five years as a launch platform, the Stratofortress had launched or released 405 payloads. In addition to serving as a launch platform, it made many research flights on its own.

Retirement

NASA's Boeing NB-52B Stratofortress Mothership was retired in a ceremony at the Dryden Flight Research Center at Edwards Air Force Base on Friday December 17, 2004.

General Curtis Bedke, the commander of Edwards Air Force Base, eulogized, "Balls Eight, you have served well and faithfully for forty-nine years. You are hereby removed from active duty on 17 December 2004, the one hundred first anniversary of the first manned, powered, controlled flight at Kittyhawk, and placed in the inactive reserves. You join an elite and honorable cadre, whose membership is significantly enhanced by your addition. Thank you, Balls Eight."

The NB-52B was the last of its kind. At the time of its retirement, it was the only B-52 powered by J57 engines that still required JP-4 fuel, years after other military aircraft had switched to JP-8. Special batches of JP-4 had to be produced to support each mission of the last B-model Stratofortress. Despite its lengthy career and the large number of programs that it supported, it accumulated less flight time than any currently operational B-52.

The NB-52B was displayed once more at the Edwards Air Force Base Open House in October 2005 and is now on display near the North Gate of Edwards Air Force Base.

The view of the right wing and X-15 pylon looking out of the Plexiglas astrodome on February 7, 2005. (Brian Lockett)

The NB-52B on the flightline on February 7, 2005. (Brian Lockett)

Boeing NB-52H Stratofortress, 61-0025 has been transferred to NASA to serve in the role of launch plane. It may launch the hypersonic X-51 Waverider. (Brian Lockett)

The narrow tubing delivers oil into the exhaust of the jet engines to generate a trail of smoke for tracking the NB-52B. February 7, 2005. (Brian Lockett)

The pilots' control panel of the NB-52B on February 7, 2005. The pilot's launch control panel is located at lower left, just below the left control yoke. (Brian Lockett)

The pilot's launch control panel of the NB-52B on February 7, 2005. A different emergency launch handle has been installed since 1996. (Brian Lockett)

The launch panel operator's station on February 7, 2005. (Brian Lockett)

The view forward from the Launch Panel Operator's station on February 7, 2005. (Brian Lockett)

The NB-52B made its last appearance at the Edwards Air Force Base Open hopuse in October 2005. (Brian Lockett)

The NB-52B is now displayed just outside the north gate into Edwards Air Force Base. (Brian Lockett)

References

1 Rotundo, Lewis; Into the Unknown - The X-1 Story; Smithsonian Institution Press; Washington and London, 1994; Pages 28-30.

2 Rotundo, Lewis; Into the Unknown - The X-1 Story; Smithsonian Institution Press; Washington and London, 1994; Pages 34-35.

3 Supersonic Flight - Breaking the Sound Barrier and Beyond; Brassey's; London, Washington, 1997; Pages 86-88.

4 Supersonic Flight - Breaking the Sound Barrier and Beyond; Brassey's; London, Washington, 1997; Pages 91-93.

5 Supersonic Flight - Breaking the Sound Barrier and Beyond; Brassey's; London, Washington, 1997; Pages 101-102.

6 Supersonic Flight - Breaking the Sound Barrier and Beyond; Brassey's; London, Washington, 1997; Pages 160-161.

7 Supersonic Flight - Breaking the Sound Barrier and Beyond; Brassey's; London, Washington, 1997; Pages 163-164.

8 Everest, Lt. Col. Frank K.; The Fastest Man Alive - The personal life story of the world-famous American rocket plane pilot; E. P. Dutton and Company, Inc., New York, 1958; Page 190.

9 First Landing of Bell X-2 Research Airplane; NACA Research memorandum RM L52I11

10 Crossfield, A. Scott with Blair, Clay Jr.; Always Another Dawn - The Story of a Rocket Test Pilot; The World Publishing Company, Cleveland, Ohio, 1960; Page 143.

11 NAA Report No. NA-59-336-36; Flight Test Progress Report No. 36, for week ending November 7, 1959 for X-15 Research Airplane; North American Aviation, Inc.

12 Crossfield, A. Scott with Blair, Clay Jr.; Always Another Dawn - The Story of a Rocket Test Pilot; The World Publishing Company, Cleveland, Ohio, 1960; Pages 222-223.

13 Crossfield, A. Scott with Blair, Clay Jr.; Always Another Dawn - The Story of a Rocket Test Pilot; The World Publishing Company, Cleveland, Ohio, 1960; Pages 265-266.

14 Contract Number AF33(038)-21096; July 3, 1958; Boeing Airplane Company

15 Contract Number AF34(601)-5252; December 12, 1958; Boeing Airplane Company

16 Letters 6517th Flight Line Maintenance Squadron to FTEA, January 2, 1959 and February 2, 1959

17 Derivation of the Weight and C.G. for Modified B-52B Airplane; dated 17 July 1959; North American Aviation

18 Flight Evaluation of the B-52 Carrier Aircraft for the X-15; AFFTC-TR-60-33; September 1960; Air Force Flight Test Center

19 Dryden X-Press, October 15, 1999

20 Crossfield, A. Scott with Blair, Clay Jr.; Always Another Dawn - The Story of a Rocket Test Pilot; The World Publishing Company, Cleveland, Ohio, 1960; Pages 308-309.

21 North American Aviation Report No. NA-58-824B, Section IV Inspection Requirements

22 Flight Evaluation of the B-52 Carrier Aircraft for the X-15; AFFTC-TR-60-33; September 1960; Air Force Flight Test Center

23 Memo: Lt. Colonel Harold G. Russell to FTE, dated 16 Nov 1959; Subject: Modification to NB-52B 52-008A

24 Memo: Lt. Colonel K. C. Dudley to Richard Harer (FTFP), dated 12 Nov 1959; subject: Modification to NB-52B 52-008A

25 Crossfield, A. Scott with Blair, Clay Jr.; Always Another Dawn - The Story of a Rocket Test Pilot; The World Publishing Company, Cleveland, Ohio, 1960; Pages 386-387.

26 Crossfield, A. Scott with Blair, Clay Jr.; Always Another Dawn - The Story of a Rocket Test Pilot; The World Publishing Company, Cleveland, Ohio, 1960; Page 389.

27 Crossfield, A. Scott, Always Another Dawn: The Story of a Rocket Test Pilot, World Publishing Company, 1960, Page 390.

28 Modification Control M-60-520Z, dated 11 Feb 1960

29 Significant Historical Data document; NB-52B 52-008; mod M-60-520Z

30 Memo: H. W. Berkowitz to FTEBB, dated 22 Feb 1960

31 Crossfield, A. Scott, Always Another Dawn: The Story of a Rocket Test Pilot, World Publishing Company, 1960, Page 391.

32 Crossfield, A. Scott, Always Another Dawn: The Story of a Rocket Test Pilot, World Publishing Company, 1960, Page 392.

33 Flight Planning and Conduct of the X-15 Envelope Expansion with Interim XLR-11 Rocket Engines; Technical Documentary Report No. 62-27; April 1962; Air Force Flight Test Center, Edwards Air Force Base, California

34 Crossfield, A. Scott, Always Another Dawn: The Story of a Rocket Test Pilot, World Publishing Company, 1960, Page 393.

35 Letter: Richard J. Harer, X-15 Project Officer to Mr. McGuire (FTFES), dated 8 Apr 1960; Subject: Work Request on B-52 Carrier Aircraft

36 Dryden Diary Entries dated November 4, 1960

37 Letter: Major John R. Lyall, Photography Branch Chief to Mr. Baldridge (FTFSF), dated 23 Jan 1961; Subject: Modification of Camera Mounts

38 Flight Test Coordination Memo: Major Kenneth K. Lewis to FTEQ, dated 30 March 1961;

39 Letter: to Mr. McGuire (FTFES), dated 8 Apr 1961; Subject: Modifications Required for Satisfactory APN-81 Doppler Radar Operations in X-15 Mother Aircraft

40 Modification Control No. M-1-F-187Z, dated 23 Oct 1961

41 Modification Control No. M-1-F-194Z, dated 12 Jun 1961

42 Memo: Colonel Clayton Peterson to Mr. Nagel (FTFES), undated; subject: Modification of X-15/APN-81 Doppler Radar for Ground Check Dummy Loads.
43 Modification Control No. M-1-J-421Z, dated 31 Oct 1961
44 Operating Instructions for dummy load - antenna switch, undated
45 Memo: Colonel Clayton Peterson to Captain Smith (FTFES), undated; subject: Addition to Class II Modification for B-52's 52-003 and 52-008
46 Flight Planning and Conduct of the X-15A-2 Envelope Expansion Program; Technology Document No. 69-4; July 1969; Air Force Flight Test Center, Edwards Air Force Base, California
47 Dana, William "Bill"; Interview with author; February 7, 2005
48 X-15 Status Report No. 64-2; Office of the Director, Flight Research Center to NASA Headquarters; dated January 16, 1964
49 X-15 Status Report No. 64-4; Office of the Director, Flight Research Center to NASA Headquarters; dated February 24, 1964
50 X-15 Status Report No. 64-3; Office of the Director, Flight Research Center to NASA Headquarters; dated February 3, 1964
51 Flight Planning and Conduct of the X-15A-2 Envelope Expansion Program; Technology Document No. 69-4; July 1969; Air Force Flight Test Center, Edwards Air Force Base, California
52 Flight Planning and Conduct of the X-15A-2 Envelope Expansion Program; Technology Document No. 69-4; July 1969; Air Force Flight Test Center, Edwards Air Force Base, California
53 X-15 Status Report No. 67-2; Office of the Director, Flight Research Center to NASA Headquarters; dated February 3, 1967
54 Legacy of the Lifting Body; Page 59; N000-0418-01; March 1995; Computer Sciences Corporation; Edwards Flight Test Center
55 Legacy of the Lifting Body; Page 43; N000-0418-01; March 1995; Computer Sciences Corporation; Edwards Flight Test Center
56 Legacy of the Lifting Body; Page 59; N000-0418-01; March 1995; Computer Sciences Corporation; Edwards Flight Test Center
57 Gentry, Jerauld, Colonel, USAF, (Ret.); Interview with Skip Stagg; Page 20; September 23, 1994; Lancaster, California; transcript LBOH-9
58 Gentry, Jerauld, Colonel, USAF, (Ret.); Interview with Skip Stagg; Page 22; September 23, 1994; Lancaster, California; transcript LBOH-9
59 X-15 Status Report No. 66-2; Office of the Director, Flight Research Center to NASA Headquarters; dated February 2, 1966
60 X-15 Status Report No. 66-5; Office of the Director, Flight Research Center to NASA Headquarters; dated May 4, 1966
61 Legacy of the Lifting Body; Heavy Weight Lifting Body Flight Log; Page 167; N000-0418-01; March 1995; Computer Sciences Corporation; Edwards Flight Test Center
62 X-15 Status Report No. 66-7; Office of the Director, Flight Research Center to NASA Headquarters; dated July 12, 1966
63 Legacy of the Lifting Body; Page 50; N000-0418-01; March 1995; Computer Sciences Corporation; Edwards Flight Test Center
64 X-15 Semi-Annual Status Report No. 6; Office of the Director, Flight Research Center to NASA Headquarters; dated November 1, 1966
65 Legacy of the Lifting Body; Page 148; N000-0418-01; March 1995; Computer Sciences Corporation; Edwards Flight Test Center
66 Painter, Weneth D.; Interview with Skip Stagg; August 15, 1994; Lancaster, California; transcript LBOH-3
67 Legacy of the Lifting Body; Page 41; N000-0418-01; March 1995; Computer Sciences Corporation; Edwards Flight Test Center
68 Dryden Diary Entry dated May 5, 1967
69 Peterson, Bruce: Interview with Skip Stagg; August 9, 1994; Lancaster, California; transcript LBOH-1
70 X-15 mission 1-73-126 post-flight pilot report; June 29, 1967
71 X-15 Status Report No. 67-7; Office of the Director, Flight Research Center to NASA Headquarters; dated July 6, 1967
72 Legacy of the Lifting Body; Page 99; N000-0418-01; March 1995; Computer Sciences Corporation; Edwards Flight Test Center
73 Significant Historical Data document; NB-52B 52-008; MOD M-7-G-107Z
74 Modification Control No. M-8-02-033, dated 2/20/68
75 Legacy of the Lifting Body; Page 61; N000-0418-01; March 1995; Computer Sciences Corporation; Edwards Flight Test Center
76 Gentry, Jerauld, Colonel, USAF, (Ret.); Interview with Skip Stagg; Page 33; September 23, 1994; Lancaster, California; transcript LBOH-9
77 Modification Control No. M-9-01-010, dated 26 Aug 1968
78 Significant Historical Data document; NB-52B 52-008; MOD M-9-A-010Z
79 Legacy of the Lifting Body; Page 68; N000-0418-01; March 1995; Computer Sciences Corporation; Edwards Flight Test Center
80 Significant Historical Data document; NB-52B 52-008;

MOD M-9-H-177Z
81 Terminal Area Energy Management, Approach and Landing Investigation for Maneuvering Reentry Vehicles using F-111A and NB-52B Aircraft; B. Lyle Schofield, David F. Richardson, Peter C. Hoag; NASA FTC-TD-70-2; June 1970
82 Legacy of the Lifting Body; Page 75; N000-0418-01; March 1995; Computer Sciences Corporation; Edwards Flight Test Center
83 Legacy of the Lifting Body; Page 123; N000-0418-01; March 1995; Computer Sciences Corporation; Edwards Flight Test Center
84 Layton, Garrison P.; A New Experimental Flight Research Technique: the Remotely Piloted Airplane; NASA Flight Research Center, Edwards, California; June 1975
85 Lockenour, Jerry L and Layton, Garrison P.; RPRV Research: Focus on HiMAT; Astronautics & Aeronautics, April 1976
86 Layton, Garrison P.; A New Experimental Flight Research Technique: the Remotely Piloted Airplane; NASA Flight Research Center, Edwards, California; June 1975
87 Technical Write-up for X-24B Runway Landing; dated August 1975
88 Legacy of the Lifting Body; Page 81; N000-0418-01; March 1995; Computer Sciences Corporation; Edwards Flight Test Center
89 Significant Historical Data document; NB-52B 52-008; W/O 12-30-19
90 Significant Historical Data document; NB-52B 52-008; W/O 35-4-13
91 Significant Historical Data document; NB-52B 52-008; W/O 35-4-15
92 Significant Historical Data document; NB-52B 52-008; W/O 35-4-16
93 Significant Historical Data document; NB-52B 52-008; W/O 31-3-001/002
94 Significant Historical Data document; NB-52B 52-008; W/O 31-3-004
95 Significant Historical Data document; NB-52B 52-008; W/O 31-3-012/014
96 Significant Historical Data document; NB-52B 52-008; W/O 31-33-016
97 Significant Historical Data document; NB-52B 52-008; W/O 30-2-006
98 Significant Historical Data document; NB-52B 52-008; W/O 12-30-060
99 Significant Historical Data document; NB-52B 52-008; W/O 35-10-001
100 Significant Historical Data document; NB-52B 52-008; W/O 35-10-012
101 Layton, Garrison P.; A New Experimental Flight Research Technique: the Remotely Piloted Airplane; NASA Flight Research Center, Edwards, California; June 1975
102 Significant Historical Data document; NB-52B 52-008; W/O 35-3-013
103 Significant Historical Data document; NB-52B 52-008; W/O 35-3-018
104 Significant Historical Data document; NB-52B 52-008; W/O 35-3-011/017
105 Significant Historical Data document; NB-52B 52-008; W/O 35-10-005/009
106 Significant Historical Data document; NB-52B 52-008; W/O 35-10-011
107 Significant Historical Data document; NB-52B 52-008; W/O 35-07-033
108 Significant Historical Data document; NB-52B 52-008; W/O 35-07-034
109 Significant Historical Data document; NB-52B 52-008; W/O 35-07-038
110 Significant Historical Data document; NB-52B 52-008; W/O 35-07-046
111 Significant Historical Data document; NB-52B 52-008; W/O 35-07-076
112 Significant Historical Data document; NB-52B 52-008; W/O 35-2-002/003/004
113 Significant Historical Data document; NB-52B 52-008; W/O 12-5-008
114 Lockenour, Jerry L and Layton, Garrison P.; RPRV Research: Focus on HiMAT; Astronautics & Aeronautics, April 1976
115 Role of Research Aircraft in Technology Development; Kenneth J. Szalai; NASA TM-85913; 1984
116 Testing of a New Recovery Parachute System for the F-111 Aircraft Crew Escape Module; Donald W. Johnson; Sandia National Laboratories SAND--89-0410C
117 Testing of a New Recovery Parachute System for the F-111 Aircraft Crew Escape Module; Donald W. Johnson; Sandia National Laboratories SAND--89-0410C
118 Dryden X-Press, October 15, 1999

NB-52B Crew Members and NASA Crew Chiefs

Pilots and Co-pilots

Allavie, John E. "Jack"
Andonian, Harry
Archer, Harry M.
Batteas, Frank W.
Bedke, Curtis M.
Bement, Russell P.
Benefield, Thomas D. "Doug"
Best, Albert H. III
Bock, Charles C., Jr.
Bowline, Jerry D.
Branch, Irving "Twig"
Buker, Russell S.
Burgenheim, Miles E.
Butchart, Stanley P.
Claxton, Richard W.
Cole, Frank E.
Cotton, Joseph P.
Cross, Carl S.
Doryland, Charles J.
Fiedler, Frederick A.
Fisher, Allen
Fornell, Gordon E.
Franzen, Ronald V.
Fritz, Nicholas H., Jr.
Fullerton, C. Gordon
Fulton, Fitzhugh "Fitz"
Goodwin, George W. III
Harper, Earle D.
Heaton, Robert R.
Hinds, Bruce J., Jr.
Hines, Joseph V.
Higgins, Thomas
Hood, Robert C.
Ishmael, Stephen

Pilots and Co-pilots Continued

Jones, Gayland E. "Gay"
King, Michael
Kuyk, Charles F. G., Jr.
Larson, George W.
LeBeau, Thomas J., Jr.
Lewis, Kenneth K.
Loewe, William R.
Luck, George E.
Lyddane, George H.
Lynch, George P., Jr.
Mallick, Donald R.
Manke, John A.
McMurtry, Thomas C.
McDowell, Edward D.
Miller, John
Morgan, Richard L.Mosley, Robert L.
Neubert, Axel P.
O'Hara, Brian
Purifoy, Dana D.
Ranz, Dale M.
Reschke, William G., Jr.
Riedenauer, Robert L.
Schneider, Edward T. "Fast Eddie"
Scobee, Francis R. "Dick"
Skabo, Paul S.
Small, John T., Jr.
Smith, Rogers E.
Smolka, James W. "Smoke"
Snyder, Cecil O.
Stroup, Floyd B.
Sturmthal, Emil "Ted"
Townsend, Guy M.
Vanderhorst, Daniel R.
Wood, William E., Jr.

Launch Panel Operators

Anderson, Brian (X-38)
Berkowitz, William "Bill" (X-15)
Butchart, Stanley P. (X-15)
Currie, Alan (X-38)
Dustin, Allen F. (X-15)
Fox, Jeff (X-38)
Henry, Jerry (F-15 SRV, DAST, HiMAT)
Horton, Victor W. (HL-10, M2-F2, M2-F3, X-24A, X-24B, SRB PTV)
McAllister, Dave (X-43A)
Minnick, Brian (X-43A)
Moise, John W. "Jack" (X-15)
Muratore, John (X-38)
Neal, Brad (Pegasus, X-43A)
Obrien, Harold (F-15 SRV, HiMAT)
Peterson, Bruce A. (X-15)
Pomeroy, John (X-43A)
Reed, Don (X-38)
Richards, Bob (Pegasus)
Russell, John "Jack" (X-15, HL-10, M2-F3, X-24A, X-24B, F-15 SRV)
Sanchez, Merri J. (X-38)
Young, Ray O. (X-24B, F-15 SRV, Firebee, DAST, HiMAT, F-111 PTV)

NASA Crew Chiefs (1976-2004)

Glynn H. Hall
Charles D. Guilinger
Daniel J. Bain
Michael D. Bondy
Kenneth Wilson

Research Vehicle Pilots

X-15 Pilot	First flight	Last flight	Number of flights Total	From NB-52B
A. Scott Crossfield	June 8, 1959	December 6, 1960	14	7
Joseph "Joe" Walker	March 25, 1960	August 22, 1963	25	12
Robert M. "Bob" White	April 30, 1960	December 14, 1962	16	8
Forrest S. "Pete" Petersen	September 23, 1960	January 10, 1962	5	3
John B. "Jack" McKay	October 28, 1960	September 8, 1966	29	17
Robert A. "Bob" Rushworth	November 4, 1960	July 1, 1966	34	22
Neil A. Armstrong	November 30, 1960	July 26, 1962	7	3
Joseph H. "Joe" Engle	October 7, 1963	October 14, 1965	16	7
Milton O. "Milt" Thompson	October 29, 1963	August 25, 1965	14	8
William J. "Pete" Knight	September 30, 1965	September 13, 1968	16	10
William H. "Bill" Dana	November 4, 1965	October 24, 1968	16	8
Michael J. "Mike" Adams	October 6, 1966	November 15, 1967	6	1
M2-F2 Pilot				
Milton O. "Milt" Thompson	July 12, 1966	September 2, 1966	5	3
Bruce A. Peterson	September 13, 1966	May 10, 1967	3	3
Donald M. Sorlie	September 20, 1966	October 5, 1966	3	3
Jerauld R. "Jerry" Gentry	October 12, 1966	May 2, 1967	5	4
HL-10 Pilot				
Bruce A. Peterson	December 12, 1966	December 12, 1966	1	1
Jerauld R. "Jerry" Gentry	March 15, 1968	December 9, 1968	9	5
John A. Manke	May 28, 1968	September 18, 1969	10	6
William H. "Bill" Dana	April 25, 1969	February 27, 1970	9	7
Peter C. Hoag	June 6, 1969	July 17, 1970	8	7
X-24A Pilot				
Jerauld R. "Jerry" Gentry	April 17, 1969	November 20, 1970	13	11
John A. Manke	October 22, 1969	June 4, 1971	12	12
Cecil W. Powell	February 4, 1971	May 12, 1971	3	3
M2-F3 Pilot				
William H. "Bill" Dana	June 2, 1970	December 13, 1972	19	19
Jerauld R. "Jerry" Gentry	February 9, 1971	February 9, 1971	1	1
John A. Manke	October 19, 1972	December 20, 1972	4	4
Cecil W. Powell	November 9, 1972	December 6, 1972	3	3
X-24B Pilot				
John A. Manke	August 1, 1973	August 5, 1975	16	16
Michael V. Love	October 4, 1973	August 20, 1975	12	12
William H. "Bill" Dana	September 9, 1975	September 23, 1975	2	2
Einar Enevoldson	October 9, 1975	November 12, 1975	2	2
Francis R. "Dick" Scobee	October 21, 1975	November 19, 1975	2	2
Thomas C. McMurtry	November 3, 1975	November 26, 1975	2	2

Remotely Piloted Research Vehicle Pilots

HiMAT Pilot	First flight	Last flight	Number of flights	
			Total	From NB-52B
William H. "Bill" Dana				
Steven Ishmael	December 21, 1981	July 23, 1982	7	7
Einar Enevoldson	October 13, 1982	January 12, 1983	3	3
F-15 RPRV/STV Pilot				
Einar Enevoldson			>15	>15
John A. Manke			>4	>4
Mike Swann			>6	>6
Firebee II Pilot				
Thomas McMurtry			1	1
DAST Pilot				
Thomas McMurtry			2	2

Key to back cover composite image

1. NB-52B carrying North American Rockwell Highly Maneuverable Aircraft Technology (HiMAT) drone, 1980; NASA photo ECN-14283
2. NB-52A carrying Northrop HL-10 lifting body with Douglas F5D Skylancer chase, 1968; Photo via Bob Dorr
3. NB-52B carrying North American X-15-3 with Northrop T-38A Talon chase, 1965; Air Force Flight Test Center History Office, 1965
4. NB-52B launching Martin-Marrietta X-24B, 1975; Photo via Bob Dorr
5. NB-52B carrying F-15 Remotely Piloted Research Vehicle/Spin Test Vehicle (RPRV/STV), 1973; NASA photo ECN3804
6. NB-52B launching Orbital Sciences Pegasus booster, 1990; NASA Photo EC90-0111-2
7. NB-52B carrying North American X-15A-2, 1967; Air Force Flight Test Center History Office
8. NB-52B dropping F-111 cockpit Parachute Test Device (PTD), 1988; NASA photo EC88-0008-02
9. NB-52B with a pair of J85 jets mounted on a bomb bay pallet for emissions control tests, 1995; Photographer: Brian Lockett
10. NB-52B carrying Shuttle Reusable Booster Drop Test Vehicle (SRB/DTV), 1977; NASA photo EC7429
11. NB-52B carrying Martin-Marrietta X-24B, 1973; Photo via Bob Dorr
12. NB-52B carrying Martin-Marrietta X-24A, 1970; Air Force Flight Test Center History Office
13. NB-52B carrying second Drone for Aeroelastic Structures Testing (DAST-II), 1982; NASA photo EC11693
14. NB-52B carrying Drone for Aeroelastic Structures Testing (DAST) with Lockheed F-104G Starfighter chase, 1979; NASA photo EC11693
15. NB-52B carrying X-43A Hyper-X Supersonic Combustion Ramjet, 2001; Photographer: Brian Lockett
16. NB-52B carrying X-38 Space Station Lifeboat with McDonnell-Douglas F/A-18B Hornet chase, 1997; NASA photo
17. NB-52B carrying Northrop M2-F3 lifting body with Northrop T-38A Talon chase, Photo via Bob Dorr
18. NB-52B carrying North American X-15-1, 1960; Photographer: Richard Lockett
19. NB-52B carrying X-38 Space Station Lifeboat, 2000; Photographer: Brian Lockett
20. NB-52B carrying North American X-15-1, 1960; Air Force Flight Test Center History Office
21. NB-52B carrying North American X-15-3, 1965; Air Force Flight Test Center History Office
22. NB-52B testing Space Shuttle drogue parachute, 1990; NASA photo

1
2
3
4
5
6
7
8
9
10
11
12
13
14
15
16
17
18
19
20
21
22

www.ingramcontent.com/pod-product-compliance
Ingram Content Group UK Ltd.
Pitfield, Milton Keynes, MK11 3LW, UK
UKHW060016051026
15591UKWH00002B/230
9780578030005